PENGUIN BOOKS

THE TELLINGTON TTOUCH

Linda Tellington-Jones is the internationally recognized expert who created "the Tellington TTouch," a method to enhance training, behavior, healing, and communicating with animals. She is the author of sixteen books in twelve languages and eighteen videos on the application of TTouch for companion animals, horses, and humans. She is the founder of TTEAM (Tellington Equine Awareness Method and Tellington TTouch Training), located in Santa Fe, New Mexico, with certified teachers in twenty-seven countries. She lives in Kailua-Kona, Hawaii, with her husband Roland Kleger, and their Westie, Rayne.

Sybil Taylor lives in New York City. She is the author of several nonfiction books, one of which, *The Last Run* (Viking Penguin 1989), is scheduled for release as a film.

For more information on the TTouch videos and other books available, please call 1-800-854-TEAM / 505-455-2945 or fax 505-455-7233.

Also by Linda Tellington-Jones

Linda has books printed in twelve languages—
Czech, Danish, Dutch, English, French, German, Italian,
Japanese, Norwegian, Russian, Spanish, and Swedish.

Available in English:

*Getting in TTouch® with Your Puppy: A Gentle Approach
to Training and Influencing Behavior.* 2007.

*The Ultimate Horse Behavior and Training Book—Enlightened
and Revolutionary Solutions for the 21st Century.*
With Bobbie Lieberman, 2006.

*Getting in TTouch with Your Cat: A Gentle Approach
to Influencing Behavior and Health.* 2003.

*Getting in TTouch with Your Dog—An Easy, Gentle Way to Better
Health and Behavior.* With Gudrun Braun, 2001.

*Improve Your Horse's Well-Being: A Step-by-Step Guide
to TTouch and TTEAM Training.* 1999.

*Let's Ride With Linda Tellington-Jones: Fun and TTEAMwork
with your Horse or Pony.* With Andrea Pabel, 1997.

Getting in TTouch: Stand and Influence Your Horse's Personality.
With Sybil Taylor, 1995.

An Introduction to the Tellington-Jones Equine Awareness Method.
With Ursula Bruns, 1985.

Endurance and Competitive Trail Riding.
With Wentworth Tellington, 1979.

The Tellington TTouch

CARING FOR ANIMALS
WITH HEART AND HANDS

Linda Tellington-Jones
with Sybil Taylor

PENGUIN BOOKS

PENGUIN BOOKS
Published by the Penguin Group
Penguin Group (USA) Inc., 375 Hudson Street, New York, New York 10014, U.S.A.
Penguin Group (Canada), 90 Eglinton Avenue East, Suite 700, Toronto, Ontario, Canada M4P 2Y3
(a division of Pearson Penguin Canada Inc.)
Penguin Books Ltd, 80 Strand, London WC2R 0RL, England
Penguin Ireland, 25 St Stephen's Green, Dublin 2, Ireland (a division of Penguin Books Ltd)
Penguin Group (Australia), 250 Camberwell Road, Camberwell, Victoria 3124, Australia
(a division of Pearson Australia Group Pty Ltd)
Penguin Books India Pvt Ltd, 11 Community Centre, Panchsheel Park, New Delhi – 110 017, India
Penguin Group (NZ), 67 Apollo Drive, Rosedale, North Shore 0632, New Zealand
(a division of Pearson New Zealand Ltd)
Penguin Books (South Africa) (Pty) Ltd, 24 Sturdee Avenue, Rosebank, Johannesburg 2196, South Africa

Penguin Books Ltd, Registered Offices:
80 Strand, London WC2R 0RL, England

First published in the United States of America by Viking Penguin,
a division of Penguin Books USA Inc., 1992
Published in Penguin Books 1995
This edition with a new foreword by Ewald Isenbuegel published in Penguin Books 2008

1   3   5   7   9   10   8   6   4   2

Illustration on page 264 by Philip Pretty

Other illustrations by Laura Maestro. Copyright © Viking Penguin,
a member of Penguin Group (USA) Inc., 1992

Grateful acknowledgment is made for permission to reprint the following copyrighted works:
Excerpts from *Kinship with All Life* by J. Allen Boone. Copyright 1954 by Harper & Row, Publishers, Inc.
Reprinted by permission of HarperCollins Publishers.
Excerpt from "Hawk Roosting" from *Hawk in the Rain* by Ted Hughes. Copyright © 1957 by Ted Hughes.
Reprinted by permission of HarperCollins Publishers.
"Sally Forth" by Marion Copeland. By permission of the author.
Excerpts from *The Peaceable Kingdom* by John Sedgwick. Copyright © 1988 by John Sedgwick.
By permission of William Morrow & Company, Inc.

THE LIBRARY OF CONGRESS HAS CATALOGED THE HARDCOVER EDITION AS FOLLOWS:
Tellington-Jones, Linda.
The Tellington TTouch: a revolutionary natural method to train and care for your favorite animal/
Linda Tellington-Jones with Sybil Taylor.
p. cm.
ISBN 0-670-82578-6 (hc.)
ISBN 0 14 01.1728 8 (previous pbk.)
ISBN 978-0-14-311456-7 (pbk.)
1. Pets—Training.   2. Animals, Training of.   3. Human—animal communication.   4. Pets—Behavior.
5. Pets.   I. Taylor, Sybil.   II. Title.   III. Title: Tellington TTouch.
SF412.7.T45   1992
636.088'7—dc20   91-29094

Printed in the United States of America
Set in Novarese Book
Designed by Liney Li

Acknowledgments

My greatest thanks and acknowledgment go to my sister Robyn Hood. It is due to her brilliance, hard work, support, and publication of TTEAM *News International* that the TTouch has developed and spread so widely. My appreciation to Phil Pretty, my brother-in-law, for his photography, art, and so many contributions. And my love and appreciation for support to my whole family.

I give my thanks en masse to our practitioners and all who have contributed to the development of TTEAM. There simply is not enough space to thank you individually.

A few, however, I must acknowledge: Ursula Bruns, who insisted I develop TTEAM into a method and coauthored my first TTEAM book for horses, making it a household word in the German horse world; Moshe Feldenkrais, whose brilliant method for expanding human potential inspired me to see animals with new eyes; Annagret Ast, who heads TTEAM in Europe with her magical presence and exquisite style; Kate Riordan, for spreading the seeds of TTEAM in so many ways; Copper Love, for lighting the lives of all she TTouches; Christina Schwartz, for her dedicated support and terrific translations; Wolf Krober, whose support has spread TTEAM throughout Europe; Andre

Orlov, for his role in the birth of Animal Ambassadors in the USSR; Elena Petaskova, for introducing the TTouch to the Soviet equestrian teams and supporting its growth in the USSR.

My thanks to Chris Griscom, Priscilla Hoback, Wendy Coleman, Dr. Nina Khanzina, Art Goodrich and Sue Goodrich, Ann Finley, Carol Lang, Alexandra Kurland, Pam Lawlor, Gigi Coyle, Harriet Crosby, Stevi Johnson, Mary Young, Pamela Philip, and Joyce Anderson for all your gifts.

To Dr. Tom Becket and Marnie Reeder, my gratefulness for the tender loving care of Kenyon and so many contributions.

To the McCulleys, the Rushes, the Sayres, and Gray Hawn and Kids TV Express, my heartfelt thanks for all you did for Kenyon and Animal Ambassadors.

I especially want to thank Dr. Ewald Isenbugel and Menja for advice, support, and friendship over two decades.

Although I have thanked Sybil Taylor at least a thousand times over the years of writing this book, I must acknowledge publicly her dedication and patience in getting the details correct and in her relentless search for words and ways to present the material with such charm.

For Sybil and myself, I thank Sybil's daughter, Erika; Sybil's trusty sidekick, Bill Saw; and Scott and Judy Taylor for being her support team; Pam Dorman, our editor, for her careful and constructive help in piloting our book; Priscilla Hergesheimer, Sheila La Farge and Sarah Venable; and our deepest love and appreciation to our agent and friend, Reid Boates, for his expertise, encouragement, and contagious enthusiasm.

—Linda Tellington-Jones

Contents

To mother and father, precious jewels in my life, and to the spirit of all the animals—scaled, furred, feathered, and finned—whose presence in our families and in the wild contributes so much to the quality of our lives

Foreword

I have known Linda Tellington-Jones since 1971. During a ride in the mountains of Los Altos, California, we started a conversation that as time continues has grown in intensity and fascination and sets the foundation for our combined work with animals of all species, living in all countries. We aim to understand animals as beings, their behavior, the important role they play when interacting with humans—far exceeding their commercial value—as partner and ambassador for understanding and introducing us to the connection with nature.

Our aim is to enhance the fascinating possibilities of connecting with animals, not just verbally, but also making a physical connection to meet the animal and influence its well-being while at the same time receiving a loving and joyous answer.

People that are close to nature, such as Aboriginals, always perceived the interspecies connection between humans and the animals they encountered to be as normal as breathing and talking before they were indoctrinated with our culture and religion. For people today this intuitive connection is almost impossible to find.

Linda still possesses this ancient ability. She has turned her amazing connection to animals into a teachable method that lovingly enhances and influences the individual, eliminates stress and fear and elicits the animal's own body to create therapeutic effects. Researchers who work with humans and animals have learned that not everything can be explained rationally, especially when considering the connection between illness and emotions. The longer I have worked with animals as a veterinarian at the Zurich zoo, the more I have set my doubts aside and realized the TTouch and Linda's connection really work.

Animals are being used as therapeutic assistants to improve human health, which is proof how much joy we would miss out on if the connection between humans and our animal friends would be lost. There is no doubt that Linda is on the right track that enhances the flow of emotion and connection between animals and humans in this well written, easy to follow book.

Surpassing therapeutic measures the book asks the fundamental questions about the connection to life and nature and also shows us the dead end road we have unknowingly entered. For the good of humans and animals I wish that this book will reach readers around the globe.

Prof. Dr. Ewald Isenbuegel
Zurich, 2008

PART I

Introducing *The* Tellington TTouch

1

Discovering the TTouch: Genesis and Development

One of my early memories is that of standing in our family kitchen in Edmonton, Canada, in the dawn light and hearing my grandfather's rabbit go thumping up the stairs to wake him. Every morning it was the same thing—the rabbit would hop up on the bed and rummage around under my grandfather's beard until he finally opened his eyes. "All right, all right," Grandpa would say grumpily.

I also remember my mother very firmly forbidding the cat to catch any mice in the house, and I remember, too, that very same cat having kittens and generously mothering a brood of ducks right along with her own litter. My father had come in from the marshes one day and produced several duck eggs from the pockets of his jacket. Quickly, my mother slipped them in under the cat who was dreaming warm maternal dreams curled in her basket with her newborns.

After several days, a clutch of wispy ducklings emerged, adding yet another ingredient to the mix of kids, cats, dogs, and birds already whirling around in the big clapboard house with the green shutters.

I heard exciting stories of the days when my other grandfather, Will Caywood, worked in Russia as the leading trainer for the one-thousand-horse stable of Czar Nicholas II.

Will used to say that to keep a horse in fine condition and working well, you had to put your hands all over the horse's whole body every day. Grandfather also confided to me that his horses spoke to him and told him who was going to win at the races.

In those early days in Edmonton, animals were not only members of the family and characters in their own right, but they were also such a natural part of life that a world without their everyday presence was unimaginable. The cows and calves of my father's dairy and pig farm gave us our living, and my pony Trixie was not only my companion but also my only means of transport, taking me on the five-mile round-trip to school through the wintry Canadian snows and the mud and greening farmlands of spring.

My three brothers and two sisters were all up on the backs of horses by the time they were four years old, and with our parents we rode in the local horse shows. Our family name was Hood and we were playfully referred to as "The Riding Hoods."

Life with the animals was an unself-conscious web of mutual support, survival, and pleasure. It was only later, when I moved out into the world, that I realized how far we humans have come from that connectedness. It seemed to me that we were distancing ourselves with increasingly dangerous results from our common bond with other species, from our old intuitive recognition that we are one with our fellow creatures, the same stuff of life, cell by cell, molecule by miraculous molecule.

Over a period of time I realized that a big part of the problem is our concept of animals as "lesser beings." As such, even when we love them, we generally feel we must be dominant. Often, in training them to our purposes, we don't consider them individual creatures capable of a surprising range of intellect and emotion. Rather, we oftentimes believe that animals learn only through repetition and force.

My first inkling that there could be different and much richer possibilities for communication with animals came, as so much of my good fortune has come, through horses.

By the time I was eleven years old, horses had become my life. Every day after school I sped to the stable of a gifted and well-known trainer-instructor named Alice Greaves. I learned eagerly from her, sharing her passionate interest in all things equestrian, and riding and training a variety of horses, mostly thoroughbreds, from jumpers to ex-racehorses. Alice was a masterful horsewoman. I can still picture her trick riding, her long hair flowing only inches from the ground as she hung upside down off the back of a galloping horse.

One of my big chores was "working out" the young, green horses. These youngsters would buck at the drop of a hat, jumping at whatever startled them. Alice had taught me the traditional method of breaking them, or "bucking them out," which meant dominating them into submission. I clamped myself like a burr onto their heaving backs and wore them into compliance with bit, bridle, and crop. The idea was (and to a large extent still is) that horses are by nature "dumb" and "stubborn" and the only way they can learn is through recognizing the superior will of humans.

The fact is that while this method produces a seemingly cooperative animal, the relationship isn't truly based on cooperation; it's based on fear. The fear creates an underlying physical tension in the horse that can burst out as aggressive or "stubborn" behavior, for which he is again punished, perpetuating an unfortunate cycle. But that's something I was to discover only several years later.

With Alice back in Edmonton I was being bucked off more difficult horses than I could count and wishing there was some other way. One evening, as I was riding home through the lengthening shadows of a late summer twilight, an old man surprised me by hobbling out of a driveway as I came trotting by. He was leaning on a cane and calling out for me to stop. "I've been watching you riding for a long time," he said, "and I have a present for you."

He reached up and handed me a worn, blue-covered book. We talked for a while and the old man told me he had been a cavalry officer in the Spanish-American War. I can no longer remember the title of the book, but it was an illustrated guide to "ground driving" as a method for breaking horses to saddle without bucking. The theory was that instead of forcibly compelling a terrified horse to accept the sudden unaccustomed weight of a rider, the trainer should work in stages, beginning by driving the horse from a position on the ground. I was fascinated and hoped I would see the old man again to thank him, but I never did.

Eager to experiment with the book's methods, I immediately offered to train a neighbor's two-year-old mare, a sixteen-hand black thoroughbred with a white blaze and two white socks. Just as shown in the book, I put a light, English saddle on her with the girth done up loosely, and then ran two long lines through the stirrups to her bit.

Using these lines as reins I ran along behind her, first at a walk, and then trotting her in a circle in both directions.

After she was completely comfortable with the feel of reins and saddle, I had her stand quietly beside a straw bale while I repeatedly got on and off the saddle until she no longer flinched. After a week we were ready—I asked her to move forward and off we went, as calm as could be, for a peacefully uneventful ride—no bucking, no shouting, no sweating, no fear.

As it turned out, the old man's blue book was a door to the future. That first experience with ground driving became the foundation for the Tellington Method, a technique for training young horses that many amateurs and professionals are now using. But in the surprising harmony and pleasure of those sessions with the black mare, I discovered more than technique alone. I received my first lesson in understanding that cooperation is not only the most joyous and natural form of communication with animals, it is also the most effective, an understanding that was crucial to all my future discoveries in working with animals.

For me, discovery is like a marvelous puzzle. First there are only a few pieces of information and then gradually more and more appear until suddenly a picture begins to emerge. The pieces come together when I let my intuition guide me, that insistent little inner voice that speaks the truth and at the same time is so hard to trust.

Many years ago I read that intuition is *unlearned knowledge*, and that still makes sense to me. It's difficult, however, in our rationally oriented culture, to put faith in insights that whisper from inside ourselves, or to trust as true and reliable inner knowledge that we arrive at inexplicably.

"How can I trust something that's just a feeling?" we say uneasily.

I remember an early encounter with this voice. I was sixteen and had just won first place in a horse show. Afterward, when my mother came back to the stable to congratulate me, she brought along a stranger, a military-looking man in his mid-thirties with a short, brush haircut and a wiry body. After sitting next to her in the spectator section, and finding out she was my mother, he had insisted on meeting me. As I looked into his vivid blue eyes, I felt something like a shock run through my body. It wasn't like the thrill of love at first sight, but more an instantaneous recognition, like a voice saying "There" and the sense that I had just met someone of enormous significance in my life.

Wentworth Tellington was a Hemingway kind of man. Inventor, engineer, explorer, cavalry officer, scholar, and teacher of mechanical drawing and polo at West Point, he also had a deep and serious interest in horses. He became my friend and mentor, and three years later I married him. As it turned out, my intuition had been totally correct—our relationship was grounded in the fact that he was an inspiring teacher to me. Together we opened the Pacific Coast Equestrian Research Farm in Los Osos, California. The farm blossomed into both an organization conducting clinical research for the improvement of equine performance and care as well as the Pacific Coast School of Horsemanship, which by the 1970s had won an international reputation, research grants from several governments, and had students who were graduate instructors from thirty-six states and nine countries.

In 1970 it became clear that it was time for Went and me to follow separate trails. We parted and for several years I ran the school myself.

Despite the end of my marriage, life was going very well. I was thirty-six years old and had spent twenty-five of those years captivated by horses—riding, showing, training, and researching them, writing articles and books about them, and teaching others what I'd learned. Every morning I stepped out to a beautiful view of oak-covered foothills rising in steep waves to the Sierra Nevada mountains. Fascinating people from all over the world came to visit me. I had a wonderful staff and caring friends. Outwardly, it was a perfect life.

Yet over a period of time I had been having increasingly troubling flashes of insight that something was wrong with this picture. One day I trucked fifteen students and a van-load of horses up Mount Tamalpais for a horse show. As we rode around the ring in 105-degree heat encased in formal hats, jackets, and boots, I looked down below to San Francisco Bay where sailboats scudded freely before the wind. What am I doing, my inner voice said, riding around this little ring for a ribbon and a trophy. Is this what I was born for?

I took time off. Sitting silently day after day in a large garden retreat, I began to feel the vibrancy of life all around me, began to hear with a new ear the nonverbal speech of flowers, grass, and trees, of insects, birds, and small animals rustling in the bushes.

I was preoccupied with communication. In the sixties I had developed "Riding with Intelligence," a technique that stressed human-equine partnership and understanding instead of the use of force, but I continued to see a discouraging lack of communication between humans and animals.

Finally it dawned on me—what I really wanted to do in life was much broader than equine education. What I truly wanted was to somehow open people to a deeper connection with animals and to their tremendous importance in our lives.

It sounded good, but naturally I didn't have the slightest idea how to go about it. Should I start with a course in animal-related studies? I went to talk with the dean of adult education at the University of Santa Cruz where I had been teaching a course in horse psychology and management.

"Linda," he said, "what you're looking for isn't taught in a university. Just go out into the world and you'll either find it or it will find you."

That was March of 1974. In April I took the leap and decided to close up shop and travel. I sold sixty-five horses, closed the doors of my school, and left California for Europe. Throughout my life my every move had been dictated by schedules and responsibility. Now here I was floating free, just following wherever my intuition led me and waiting for the next door to open. It was exhilarating, like following the clues on a treasure hunt.

It didn't take long for my new direction to appear, although I didn't recognize it at the time. Two major encounters occurred that were to place me firmly on the path toward my new work, the first with a desperately lonely female gorilla and the second with a brilliant Israeli physicist.

Through my work with horses over the years I had made many unusual and dear friends all over the world and after spending a few weeks riding in Ireland I decided to visit one of them. He was Ewald Isenbugel, the curly-haired young director of the veterinary department for the Zurich Zoo in Switzerland. Ewald had become interested in my training methods, and he and his fiancée Menja had visited me at my school. We talked many evenings away, sitting on my patio and discussing his patients, from antelopes to tigers.

And so it was that on a gray afternoon in early December I found myself standing in front of the gorilla cage in the Zurich Zoo. Ewald, who had been showing me around, had been called away.

The gorilla was alone in a twelve-by-twelve-foot enclosure, without even a view of other animals to give her a sense of companionship. In this bare and solitary confinement she was literally going out of her mind with boredom, biting quietly on her nails, picking her nose, propping her chin on her hand then restlessly shifting to another position, now and then going over to a corner to lick up what she had previously regurgitated. Somehow I just couldn't leave her, but remained standing in front of her, feeling her misery in my own bones.

I made a vow then. I promised that I wouldn't forget her or the suffering of zoo animals like her, even though I didn't yet know how I could help. To record the boredom and loneliness of her life, I snapped a photo of her every sixty seconds for an hour—sixty photographs. I have never forgotten her inspiration and keep a picture of her nearby to this day (see photo insert).

I think we all have certain objects, certain events, places, and people in our lives that are like important touchstones to us. For me, Ewald Isenbugel and the Zurich Zoo is such a touchstone. I have returned to visit many times since that first winter evening, and since then there have been many heartening changes, including a complete transformation of the gorilla enclosure.

In today's Zurich Zoo, gorillas no longer endure lonely isolation, but live together as a family group in which mothers are allowed to keep their young, a departure from many zoos' policy of taking babies away to be raised by hand. The gorilla enclosure is filled with interesting playthings and the environment is changed every day. For instance, on one day water splashes from a shower fixture; on another raisins are stuffed into holes in tree branches so that the gorillas can have fun digging them out with sticks.

In the winter of 1974, however, these changes and my own work with zoo animals were still in the future.

I had thought when I left California that I was finished with horses, but it seemed that horses, who had already brought so much to my life, were not finished with me. Several days after my encounter at the zoo, my German friend Ursula Bruns called me in Zurich and offered me an irresistible challenge—an invitation to orchestrate a riding performance for Equitana, the largest equestrian trade fair in the world.

Ursula Bruns was a tall, dynamic woman in her early fifties. She is famous in the international horse world for authoring over forty books on horses and for being an expert on Icelandic and African horses, but most of all she is known in Germany for her unorthodox approach to "free" riding (riding that is for pleasure rather than for competition). There was nothing and nobody that Ursula wouldn't take on, no rule too "sacred" to be examined. Her sponsorship of my first performance for Equitana was to usher in a whole new phase of my life.

The fair, held every two years in Essen, Germany, draws crowds of over two hundred thousand people to ten huge exhibition halls. Over eight hundred companies from thirty countries display almost every item that has anything to do with horses, from a five-dollar cake of saddle soap to five-thousand-dollar sunlamps. Every day for three hours, the arena, colorfully decorated with the flags of all nations, becomes a whirl of manes and tails as horses are shown for sale— five to ten horses every two minutes—from Russia, South America, Saudi Arabia, from all over the world. Evenings are devoted to theatrical riding exhibitions based on mythological or historical themes, lavish pageants attended by enormous audiences.

I went off into the wintry German countryside with Ursula and two other riders and trained for several months. Finally we were ready and our performance at Equitana caused a minor sensation. It was a jumping exhibition executed Native American–style, bareback without benefit of saddles or bridles. Most Europeans, used to the discipline of classical dressage, were completely unfamiliar with this type of relationship between horse and rider.

Afterward, many of the German horse magazines speculated about how it had been done, including the notion that the horses' tongues had been tied down all night to make them docile. We had used nothing in the horses' mouths except for very light, breakable strings, but it was hard for people to believe that we could achieve such results without the use of force.

After Equitana, I was asked by the German government to pilot an instructional program in "American-style" riding for a proposed official German riding center. I began traveling all over Germany to research the project, meeting people who wanted to improve their ability as riders and learn to handle their horses' behavior problems and dysfunctions. It was wonderfully stimulating and I was even learning more German than just *schritt* for trot and *pferd* for horse.

It was at this point that my second breakthrough encounter occurred. Through a friend of mine I had heard about the exciting physical therapy techniques of Dr. Moshe Feldenkrais. I sent for material on his teaching, read it, and realized that his ideas held important answers for me. This time it didn't take months to listen to my intuition. I left Germany for the summer to enroll in his twelve-week initial course at the Humanistic Psychology Institute in San Francisco.

A celebrated Israeli physicist, athlete, and master of the martial arts, Moshe discovered his revolutionary methods of mind-body reintegration literally by accident. He was crossing a street when he was hit by a bus and both his legs were severely crippled. Though his doctors gave him only a fifty-fifty chance of recovery even if they were to operate, Moshe was not a man to give up. He decided against surgery.

Determined to reeducate his legs and body, he worked out a program that would bypass the habitual way he had once moved and would instead utilize every conceivable alternative motion available, right down to the tiniest and subtlest flexing of a muscle. In two years he was walking again. He had discovered the theories and practices that eventually would help not only those with paralysis and functional impairments, but dancers, athletes, and anyone who wanted to reach his or her fullest physical and mental potential.

His techniques, now known throughout the world as "Awareness Through Movement" and "The Feldenkrais Method of Functional Integration," are based on the amazing fact that we use less than 10 percent of the brain cells with which we are born.

As we develop and learn to walk, talk, dance, and ride horses, habitual patterns of neural response are laid down between certain brain cells and certain muscles. For instance, if we have learned to walk in a slumped way, the body will then be programmed to continue to walk that way. If we have learned French, the muscles of the mouth will automatically and habitually be held differently for speech than if we have learned English or Swahili.

The parts of the brain that are so programmed are the only parts we use. Dr. Feldenkrais, however, developed a system of gentle, nonthreatening movements and manipulations that, because they are nonhabitual, awaken new brain cells and activate unused neural pathways. The body's stored bad habits and those patterns formed in response to tension, pain, and fear are broken, releasing the "cramped" emotional attitudes that go with them. Now, new choices are possible and with new choices comes a new ability to learn—coupled with a renewed self-image. The purpose of his work, Moshe often stated, was to give people a means to reach and fulfill their own highest potential, whether physical, emotional, or intellectual.

The Feldenkrais training was a course of learning taught over four consecutive summers at Lone Mountain College in San Francisco, in a lovely old building that had once been a convent. The spacious rooms with their polished wood floors were perfect for our large classes of sixty-five people all doing floor exercises. Though I didn't yet know it, the place would become like home; I was to spend three summer months out of every year there for the next three years. During the remainder of the year I continued to teach, giving riding workshops throughout Europe and integrating the Feldenkrais concepts into my own methods.

The first two summers at the institute were spent lying on the floor and learning the series of minute, nonhabitual movements that were the transformative base of the training. These were the infinitely subtle yet specific movements that brought new awareness to the body and activated hitherto unused brain cells.

It was Moshe's belief that before we could effectively work on others, we had to thoroughly experience exactly how this "new awareness" was brought to being in our own bodies. By the third year, with our foundation of personal experience of the movements in place, we were ready to work on each other, learning how to adapt our individual knowledge to another being and giving each other feedback.

Finally, by the fourth summer, we graduated to working with others outside of the school training sessions, albeit still under supervision. It was a fascinating time. Part of our graduate training was to observe Moshe's own practice of functional integration. All kinds of people from all over the world came to him and to his Israeli assistants. Some of them had been unable to find help for their conditions or relief from pain for years. They came with paralysis, with extreme pain in the neck, back, and hips, with all manner of dysfunction.

Sometimes it was really hard to detect what he was doing because the movement he was making would be so fine. For instance, for seven sessions he might do no more than repeat a tiny movement on the leg of a paralyzed person. There were some amazingly dramatic and poignant moments; I remember seeing a paralytic who hadn't walked in years begin to crawl after the seventh session and walk after the tenth.

Of course, throughout the entire four-year training period we worked with the practice not only during the intense learning period of the summer, but every day—by ourselves and with others, exploring the movements and honing our physical and mental grasp of the method.

In my first weeks of studying with Moshe I began to realize I had entered a brand new world. Moshe led us through the movements, interspersing these sessions with explanations of how he had created and found the work by trusting the process as it unfolded for him. Sometimes distinguished followers of his work like Margaret Mead (whom he treated for her limp) or Carl Pribram, the eminent Stanford University neurologist, came to visit and speak to our class.

Moshe was not a tall man but he had riveting presence. A master in karate, he was solidly built, with a barrel chest and a fluid way of moving in spite of his destroyed knees. Five days a week, from 10 A.M. to 4 P.M., he was there with us as we lay on the floor of the high-ceilinged rooms and moved our bodies in very slow, minute ways that at first drove me crazy. I was an athlete, used to moving purely in-stinctually, and here I was trying to control every infinitesimal shift of a muscle.

On the second day of class something Moshe said hit me like a bolt of lightning. To learn something, he told us, you don't have to repeat an experience over and over. Repetition isn't "learning," it's just performing the same motions again and again. But when you move the body in a nonhabitual way without fear—the human nervous system *has the ability to learn in only one experience.*

In *one* experience—I was amazed and fascinated. I flashed on the endless hours, days, weeks, and even years that I'd spent "longing," or teaching horses to walk, trot, and canter on command while I held them on a line. Some of the young horses I'd worked with had learned phenomenally fast, and yet instead of taking my cue from them and moving on to the next step, I had just gone on grinding away at the poor things with the same repetitive lessons. I believed what I had been taught by all the literature and conventional knowledge about horses—that if you don't use repetition they won't learn.

Moshe was talking about humans, but couldn't his principles of learning also work with horses? Horses get in trouble when they don't do what humans want, when they frustrate human demands with what appears to be resistance to learning. Why not find a way to develop nonhabitual movements that would eliminate whatever tension was causing the resistance? The horse's potential for learning would be released—perhaps even in one session.

I was very excited. Moshe had given me a key to what I had been looking for—a new approach to the problem of relationship and communication between humans and animals. I couldn't wait to try out his theory on a horse I had heard about.

She was a sixteen-year-old Arabian that Ted, a neighbor and friend in the area, owned. She had been a brood mare in Montana, had never been ridden, and was really not very interested in people. The problem was that each and every night Ted had a battle just to catch her and bring her into the barn—and he was getting pretty sick of it.

I went to see her and after Ted finally caught her, I stood there thinking. This horse needed to break a bad habit. To do that, using Moshe's theory, I would have to try and activate unused neural pathways to the brain. I would have to figure out ways of moving the mare's body in gentle patterns that she could not do for herself.

I approached her quietly and gently took hold of her ears, rotating them in slow circles in both directions. Then, holding her ears at the base, I waggled them back and forth, something I had discovered horses seem to like. Next, I pulled out the upper and then the lower lip and rotated them in both directions, simply because that was the only thing I could think of to do that was nonhabitual.

The mare was standing very still, becoming more and more relaxed. Going to her legs, I rotated them gently and moved them around in as many directions as I could, and followed up by doing the same for the tail, arching it up and then pulling and releasing it, giving a stretch all along the spine.

Moshe taught that when any one joint in the body was manipulated in an unusual or nonhabitual way, all the rest of the joints in the body would also be affected. Since there are from eighteen to twenty vertebrae in the tail, I figured that moving the tail was a perfect way to create a great deal of corresponding movement throughout the skeleton.

When I finished, after forty-five minutes, the horse's eyes were half closed. She stood with her head lowered, obviously in a state of deep relaxation. Ted and some other friends who had been watching had started out the session chatting to each other, but now they, too, had fallen into a mood of silent tranquility.

What had happened to us all? we wondered. It was almost hypnotic.

Interesting, I thought, really nice. But I went home not expecting any great miracles. The next night, however, Ted phoned with exciting news. Previously, he had always had to go out into the pasture to lure the mare into the barn for the night with bribes of grain. This time he could hardly believe his eyes. There she was, not only eager to come in, but actually standing at the gate, waiting. She walked docilely to her stall, but instead of rushing to her dinner in the feed bin as she usually did, she stood looking at Ted expectantly. "As though," he said, "she was offering me her ear for some more of that good stuff."

I was elated. All that summer I explored how what I was learning from Moshe could be applied to problem horses. Working with students' horses, I discovered that more often than not horses that were thought of as having "aggressive" or "resistant" personalities were just expressing pain, tension, or physical discomfort. As soon as the pain was relieved their "bad personalities" changed as totally as though a wand had been waved over them. That is not to say that all horses are born with gentle personalities—some are born with incapacitating tensions. They vary, of course, just as humans do.

After that summer, the early concepts for the Tellington TTouch began to evolve faster and faster, almost by themselves, as all of my experiences, methods, and intuitions, old and new, came together. It was a process that seemed to unfold as naturally as a plant when the time has come to blossom.

Word had gotten around the European equestrian world that I was developing and teaching a method of horsemanship called "Riding with Awareness," inspired by my Feldenkrais work, that was especially helpful to riders with problems. Over the next three years I divided my time between summers of studying with Feldenkrais in San Francisco and winters traveling throughout Europe, teaching small groups of riders as well as giving demonstrations of my techniques before larger and larger organizations and audiences.

I continued exploring the body work by working with the fears of both horses and riders and developing systems that woke up a horse's intelligence, like ground driving and walking through labyrinths—techniques that were somehow coming together to form a coherent method. To everyone's excitement, including mine, severe problems were often solved in several sessions. In addition, horse and rider learned to speak the same nonverbal language.

During this period Ursula Bruns once again stepped in with her encouragement and good German practicality. "You have a wonderful new method here," she said, "but it's not organized so that it can be taught by others. You need to do that." We decided to systematize my discoveries through a research project to be conducted on twenty horses for five weeks at her training center in Reken, Germany.

The group of owners came from all over Germany and their horses had a great variety of problems. One shied dangerously at any piece of paper or plastic that blew in the wind, another attacked his owner; others were flat out runaways or so nervous they were a danger to themselves and their riders. We also took on several horses with severe physical disabilities.

The point of the project was to eliminate these problems through a systematic approach to training. We set about defining a series of exercises called Tellington Equine Awareness Movements (later to evolve into TEAM) in which the owners, through the actual process of teaching their horses, found themselves also learning—the happy result being a partnership in which together both horse and rider learned the art of cooperation.

My original groundwork now had grown to comprise a whole system of training patterns in which the handler first led the horse in a variety of ways from the ground rather than from the saddle. Relieved of the physical and psychological pressure that often comes with bearing the weight of a rider, the horse was better able to learn. The difference this made in speed and willingness to learn was phenomenal.

Suddenly, as humans learned to teach by encouraging rather than dominating, training sessions were turning from confrontation to fun.

Horse and human were like two dancers, with the human as the leading partner, guiding and informing rather than coercing the horse's movements.

Another type of ground work was executed by laying poles in various ways to form a labyrinth through which the horse was then led. This checked the horse's natural, habitual movements, teaching her to focus and to pay attention, reactivating her learning ability.

The body work was a different story. It was very difficult to teach people how to work on a horse's body, and doubly hard to demonstrate the very subtle movements of the Feldenkrais method on such a large creature. What we did do was teach people how to move the horse's joints in ways that he could not do himself; how to move the animal's head and get him to lower it, which we found made a spectacular difference in trust; how to work the ears (it was only later that I was to learn the great value for all creatures in working the ears), and how slowly to rotate and gently pull the tail.

At the end of the five-week period, sixteen out of the twenty horses were well adjusted and totally free of the fears and bad habits that had brought them to us. That was wonderful, but the remaining four cases had made little significant progress and that really bothered me. I decided to go to work on them myself, concentrating especially on body work. One of the four horses responded beautifully, but I gave up on the other three, regretfully returning them to their owners.

Not every horse is suitable for the TEAM training or for the rider. Sometimes horse and rider are mismatched either in personality or in type. For instance a person who wants challenge and quick intelligence might have a solid but not very bright animal who would be more suitable for a less adventurous rider. Sometimes, too, our methods simply fail with a particular horse, and to go on pushing is neither safe nor appropriate.

There is also, however, the question of persistence. The owners of one of my "failures" refused to give up and kept on patiently applying our methods. After six months the horse had improved dramatically and later went on to win several one-hundred-mile endurance races.

I had been exploring how to utilize the body work all along, learning acupuncture theory and applying it to the horse's anatomy, but the basic principle for what was to become the TTouch came to me one day early in my studies with Moshe. I was sitting in a Mövenpick restaurant in Stuttgart reading a book that was on our reading list: *Man On His Nature*, by Sir Charles Sherrington, the Nobel Prize–winning British physician famous for his work on the nervous system.

People talked and laughed around me, my coffee and apple torte sat unnoticed. I remember reading the astonishing information that if you cut four inches out of a nerve pathway, most of the time the two ends will find their way back and join together again. How do they *know* how to do that? I thought.

They know, I read on, because each cell in the body has an intelligence all its own. When cells are in a normal state, they know their function within the body, what part of the whole they are.

Moshe's technique activated the neural pathways through manipulation, but learning this procedure took years. Also, it was very complex to teach, especially on a creature as large as a horse.

But here was Sherrington talking about the individual cell itself. What if through nonhabitual touch you could stimulate the intelligence of each cell and so turn on the corresponding brain cells like so many light bulbs? But how to do this?

This idea germinated for several years but it wasn't until 1983 that it bore fruit. By then I had finished my training with Moshe and had returned to California to teach my method, which by now had a lot of friends worldwide and a name—TEAM, for Tellington-Jones Equine Awareness Method.

Ursula Bruns and I had collaborated on a book about the method, and I was traveling widely to work with professional and Olympic dressage horses as well as amateurs. My baby sister Robyn had grown up to become a great horsewoman and the perfect partner for me. With her help TEAM had taken on a life of its own, with its own practitioners and a newsletter that also served as an information exchange. Letters poured in from around the world, warming us with success stories and case histories of what TEAM had accomplished. We were getting invaluable feedback.

But right up until the summer of 1983 I was still having a hard time figuring out how to teach people direct communication through touch, a type of intuitive *listening* with your hands that was impossible to describe.

One day at a clinic, a woman standing beside her horse, said to me in frustration, "But Linda, just exactly what am I supposed to *do* with my hands?"

"Just push the skin in a circle anywhere on the horse's body," I told her, "and breathe into the circles as you make them." It was something that came out quite spontaneously, without thought, but it turned out that the little circles allowed her to focus both her own and the horse's attention at the same time, and in a way that brought a deep sense of communion.

And so, almost unnoticed, the solution to my problem had appeared, the circular touch that was to be the foundation of TT—the circle, ancient symbol for life unending, for renewal, community, wholeness, and self.

Gradually over the years, as we worked with hundreds of horses and our successes became known, people began to inquire about other animals with problems. They asked for help with their dogs and cats, their farm animals, their exotic pets (our first, an unsociable rosy boa snake who learned the pleasures of affection), and to our excitement we saw the circular touch begin to emerge as a means of healing not only behavioral but emotional and physical symptoms.

With experience, we evolved (and continue to evolve) many different and effective ways of holding the hands and fingers and of applying varying pressures, with remarkable results.

My practice of TT and TEAM was now taking me all around the globe—not only to Europe, but also to Africa, Australia, and Russia—and everywhere I went I made friends. Love of animals was like a universal solvent that instantly melted away the barriers of language and culture. In a Moscow store, for instance, the clerk was cool and formal until she saw the Zuni fetish necklace I was wearing, decorated with dozens of tiny, colorful gemstone animals—birds, wolves, turtles, horses, bears.

Her face brightened immediately when she saw it. "How beautiful," she said, and asked me about it. The conversation then led to the fact that I was in Moscow to work with horses and other creatures. She got very excited and called several more people over, and in no time we were all jabbering away in broken Russian-English, exchanging stories with much smiling and gesturing.

In all of my travels I never did forget the distress of the lonely gorilla in the Zurich Zoo and the vow I had made to her, and at last, in 1984, I was finally and miraculously able to do something about it. My first consultation was in the spring of that year for the San Diego Zoo, where Master Zookeeper Art Goodrich asked me to advise him on a female Somalian ass who was being bitten by the male and was too nervous to treat. That was the beginning.

Since then I've been able to demonstrate the TTouch to keepers, directors, and veterinarians the world over. I've had the honor of making friends, of healing and communicating in a very special way with exotic creatures great and small; with elephants and lizards, with apes and snakes, with antelopes and tigers, wolves, bears, parrots, and cranes, with dolphins and even a snail.

Discovering the TTouch has certainly been a magical process, but TT itself is not "magical" at all. It is as practical, accessible, and real as your own two hands. Many people come to me and say, "Oh, but of course you are some kind of a wizard, a healer with special powers which you've spent years learning."

Please know that the gift for interspecies communication is there for us all, that there is nothing I do with TT that can't be done by you with your own two hands for your own animal—for your dog or cat or bird or gerbil. All it takes is your desire, your love, some trust in your intuition—and, of course, a little practice.

2

The Basic TTouch

> The power of the world works in circles . . . The sky is round, and I have heard the earth is round like a ball, and so are the stars . . . even the seasons form a great circle in their changing . . . and so it is in everything where power moves.
>
> —Black Elk as told to
> John Neihardt

The TTouch always makes me think of the rabbit hole in *Alice's Adventures in Wonderland*—it looks so innocent on the surface, yet it is an opening to a whole new magical world. You make circles holding your fingertips, your fingers, or your hand in various positions using varying pressures. Simple. But these simple circles, like the rabbit hole, are an entrance, a doorway into a whole different dimension of relationship with your animal.

I can't explain why the circles work any more than I can explain the miracle of falling in love, of dandelion seeds blowing in the wind, or of a beautiful horse in motion, yet there are a few basics that we can understand.

The cells that form each living being all share the same universal "intelligence." Every cell "knows" how to be a perfect part of a feather, a twig, a hand or a paw. It knows its function in the individual body at the same time as it knows its function in the universe.

The circles of the TTouch seem to provide a way for this "cellular intelligence" common to all life to become communication, for the cells of one being to make a direct connection with the cells of another. It's as if the TTouch communicates across species barriers like a person-to-person call in which the same language is spoken although the callers are from different countries.

Your circular "call" wakes up the cells of that other being, activating them to release stored memories of pain, "dis-ease," or the expectation of pain, and allowing them to "remember" their encoded potential for perfection.

Of course, the beauty of communication is that it is a mutual event, a *two-way* exchange, so when you touch an animal at such a direct level, both of you are affected. You feel a nonverbal connection that is new and different, a deepening current in relationship.

As I've said, I first discovered the circular TTouch while trying to teach people body work on horses. Later, as we continued to experiment, I added rhythmical breathing, a variety of hand positions, and a scale of finger pressures from one to ten. To my amazement, a person having less than an hour's instruction could often make major changes in behavior and personality in animals, as well as considerably speed up the healing of wounds, injury, or stiffness in themselves or their animals.

Over the years, these first basic elements of the TTouch were refined and organized into a core method that consists of fifteen hand positions and movements that we continue to research and expand (see page 240). We gave each TTouch an evocative animal name like "The Lick of the Cow's Tongue" or "The Clouded Leopard," finding that associating a particular TTouch with the characteristics of a particular animal made the movement much easier and more fun to learn, while also bringing people new and unusual ways of relating to the animal kingdom.

We then organized the basics of learning the TTouch into four components: mental attitude, using the hands and fingers, breath awareness, and finding the pressure scale.

MENTAL ATTITUDE

The watchword here is open-mindedness. It's very important to clear your mind of any preconceived ideas because expectations can be limiting. The best way to approach the TTouch is with a readiness for anything to happen, an attitude that I call the "gourd head" state.

The concept comes from a novel I once read about an amazing samurai in fourteenth-century Japan, a kind of knight-protector of the realm who won all of his battles in spite of the fact that he was blind. His secret when challenged was to visualize his head as a hollow gourd empty of all thoughts and images. Because nothing in his mind colored or blocked his incoming perceptions, all of his senses became super sensitive to the slightest nuance of what was happening around him—like a kind of second sight.

When you sit down to work with a creature, simply focus what you're doing instead of *trying* for connection. Seeking too hard or waiting for some kind of heightened contact will not only block your own experience but will also make your animal nervous and self-conscious.

Thinking of yourself as possessing a *special* healing quality will also limit your ability to be open to the moment. Everyone can practice the TTouch—you don't have to be a specialist or a therapist. Because you are not manipulating energy, bone, or muscle, you do not have to be proficient in anatomy or bioenergetics. You are simply working cell to cell, awakening the regenerative potential and cellular "intelligence" that resides as equally within the being you are touching as it does within yourself.

I like to compare the TTouch to what happens when one car battery gives another one a jump start: once the initial contact is made, the car that needed the boost can drive off very nicely under its own power and doesn't need the other car anymore.

HOW TO USE YOUR HANDS AND FINGERS

The foundation of the TTouch method is a circular movement called the Clouded Leopard (page 241). It is the first TTouch we teach because the techniques and principles used are basic to all the circular TTouches.

Begin by orienting yourself: visualize the face of a clock on the animal's body, about half an inch in diameter. With one hand supportively resting on the animal, take your other hand and place your fingers at six o'clock on your imaginary clock. With your fingers held in a lightly curved position like a paw, push the skin around the clock in a clockwise circle. *Maintain an even, constant pressure* all the way around again, past six, until you reach eight o'clock. At eight, pause for a second, and if the animal seems to be relaxed with the contact, bring your fingers away softly and begin again at another spot chosen at random.

In the basic TTouch, we place the circles at random as a way of keeping the animal focused. She remains in a state of attention because she is wondering where the next move will come from. Because each circle is a complete movement within itself, you can work the body in any order you wish without losing effectiveness.

It's important to make only *one circle and a quarter at a time* on any one spot. Practically everyone learning the TTouch has a natural tendency at first to keep on circling over the same place. Oddly, most people seem to favor making three consecutive circles. The only effect this has, however, is to "bliss out" both giver and receiver. Not to say that "blissing out" isn't delightful—it simply isn't what the TTouch is about.

When you make the circle, rest your wrist, thumb, and little finger lightly against the body to provide steadiness, and move your middle three fingers flexibly as one. Don't hold the joints of your fingers stiffly, but allow them to relax and move with the rotation. To see the difference in effect, try making a circle on yourself holding the fingers stiffly and then try the same circle with the joints of your fingers, particularly the first joints, softened. It's much easier with the joints mobile, isn't it?

In working the TTouch it's important to make sure that your circles are really round and that they are made in one smooth, flowing movement.

If, when you make your initial circles, you see that the animal is a little bit concerned about the contact, make the circles fairly fast, taking about one second to make the circular motion. Then, as the animal begins to trust and enjoy what's happening, as you feel him begin to enter a state of bodily "listening," slow the circles down to two seconds. To complete this slower circle, instead of simply lifting off when you reach eight, pause, push in a little, and then allow your fingers to come up in a gradual release, as though a sponge was slowly pushing them up and away from the body.

The first, more speedy circle awakens the body. The second, slower approach releases muscular tension, enhances breathing, and gives you a key to deeper communication.

BREATH AWARENESS

We all have a tendency to arrest our breathing when we concentrate, but that creates stiffness and tension and blocks a smoothly flowing connection to the world. Compare what you feel when making a circle holding your breath with what it's like when you breathe freely and rhythmically. Isn't the difference amazing?

As you make the circles, be sure that your breath keeps flowing in an even and easy way. Develop an awareness of your breathing and play with it until it feels rhythmic.

This type of breathing will aid greatly in stilling and focusing your mind, relaxing your neck and shoulders, and softening your hand. Your fingers will seem to be reaching out effortlessly on their own. Often, the animal you are working with will gradually attune his or her breathing to yours, creating a mood of heightened receptivity between you.

FINDING THE PRESSURE SCALE

One of the most common questions I hear about the TTouch is "Don't you get tired doing this work?"

"No," I say. "If you're getting tired it means you're doing something wrong." Effective pressure is related not to muscle power but to how you hold your hands and how you breathe. When you're doing these things effectively, you don't have to be a bodybuilder to exert the right pressure.

The TTouch employs a pressure scale from one to ten. To learn the measure of each number begin with number one as a guideline. First, bring your right hand up to your face, steadying the bent right elbow against your body with your other hand. (Do the reverse, of course, if you are left-handed.)

Then, placing your thumb against your cheek in order to give your hand a support, with the tip of your middle finger push the skin on your eyelid around in a circle with the lightest possible contact. (If you are wearing contact lenses do this on your forehead rather than on your eyelid.) Take your finger away and then do it again to get a sense of just how light that very lightest TTouch feels.

Next, on the fleshy part of your left forearm, make a circle using the same pressure as you did on your eyelid, and observe how little indentation you make in the skin. That lightest circular pressure is number one. Two is just slightly heavier.

To find three, repeat the process, only this time push the skin on your eyelid around in a circle as firmly as is comfortably possible. (I always emphasize the word "comfortably" and tell my students the idea isn't to pop out your eyeball.) Then, retaining the sense memory of that pressure, go once again to your forearm to see how the TTouch feels there and to confirm the depth of indentation into the muscle. Staying with your forearm now, go back to the one pressure and compare it with the three. Note the difference.

A pressure three times deeper into the muscle than a three is a number nine or ten. If these heavier pressures are done with the pads of the fingers they could hurt both doer and do-ee, whereas if you tip your fingers forward and lead with your nails you can go way into the muscle with no discomfort (see Bear TTouch and Tiger TTouch, pages 246 and 252).

You'll find yourself experimenting with pressure until you click into the one that is definitely "right" for the animal you're working with. Needless to say, small creatures call for only the lightest of pressures. Animals that are heavier or larger may be more responsive to the deeper pressures, but not always. If they have any sort of pain or inflammation in the body you may have to begin work with a two or three pressure.

While the basics I've outlined here may seem complicated on paper, they are actually surprisingly easy to learn. After a little bit of practice, you'll find yourself making the circles and blending the four elements of the TTouch naturally, intuitively, and let's not forget, pleasurably.

TTEAM AND TTOUCH

Throughout this book you will be reading references to TTEAM as well as to TTouch because these two terms are vitally interrelated. "The TTouch," or "TT," for "Tellington TTouch," refers to the system of circular movements and the principles underlying their application. After we adopted our acronym TT I was amused and intrigued when someone pointed out to me that it is also the Greek letter *pi*, a mathematical symbol used to designate the universal and unchanging relationship between the diameter of a circle and its circumference.

TEAM was the original acronym for Tellington-Jones Equine Awareness Method, the name given to the system of training and body work for horses that I spent so many years developing. After my Feldenkrais studies, as I continued to explore the specific possibilities of equine body work, the TTouch emerged. We found that while the TTouch could be used as an independent system, it also remained an integral part of the training methods we had begun to call TEAMwork. To indicate that TEAM employed the TTouch we added an extra T, making TTEAM.

Eventually, as we began to adapt the TTEAM methods for work with dogs, cats, bears, cows, and all the other myriad species, we took on an additional definition for our acronym: the Tellington-Jones Every Animal Method, a name thought up by a group of children who had taken part in a special TTEAM education program we had created for the Idaho school system.

And of course the name represents the goal of our work: to help you and your animal truly become a team, with all the communication, fun, and cooperation that the word implies.

HOW TO USE THIS BOOK

For easy reference, and in order not to overload the the general flow of this book with too much detailed instruction, we have grouped the definitions and specific explanations of various TTEAM terms and TTouches in a "How-To" chapter (Chapter 12) at the end. It's probably best to glance at this section first in order to familiarize yourself with our concepts and terms. Then, later, as you read the body of the book and gain a fuller understanding, you can go back to the "how-tos" for specific references or more detail as needed.

PART II

The
Animal
Connection

3

A Cat Can
Look at a King—
From Tiger to Tabby

Pangur Ban

I and Pangur Ban, my cat,
'Tis a like task we are at;
Hunting mice is his delight,
Hunting words I sit all night.

Better far than praise of men
'Tis to sit with book and pen;
Pangur bears me no ill will,
He too plies his simple skill.

'Tis a merry thing to see
At our tasks how glad are we,
When at home we sit and find
Entertainment to our mind.

—Ninth-century Irish verse
    translated by Robin Flower

Cats are the most recently "tamed" of our domestic animals, having thrown in their lot with humans only about three thousand years ago, compared to ten thousand years ago for dogs and five thousand for horses. There are some who say that cats have never really settled in and are still of two minds about setting up housekeeping with us.

After all, they can be so independent and haughty, so aloof. Yet at the same time, when describing someone who is kind, easygoing, and friendly, we refer to the person as a "real pussycat."

While we perceive cats to be of two minds about us, it is actually we ourselves who have always been of two minds about them. Throughout religious history the cat has been cast in a dual role, as symbol both of good and evil, light and darkness, sun and moon, as divine power and black magic.

In ancient Egypt, besides being an important god, the cat was considered so sacred that anyone who killed one was himself executed. When a household cat died, the occupants of the house shaved off their eyebrows in mourning and performed complicated funeral rites, including burying the embalmed body of their cat in a special cat necropolis.

In South America, both Aztecs and Incas worshiped the cat; in Burma sacred temple cats were believed to harbor the souls of the faithful dead; and in northern Europe the cult of Freya, the Norse goddess of love and beauty, included the honoring of cats among its rituals, and the goddess herself was said to travel in a chariot drawn by two felines.

On the "shadow" side of veneration, cats have been associated with the unconscious forces of nature and with the power of priestesses and witches from the earliest times of mythology and folklore. Ovid tells us that the moon goddess Artemis transformed herself into a cat when the gods fled from Mount Olympus into Egypt. Hecate, goddess of the underworld and the dark facet of Artemis, also changed herself into a cat. And in medieval Europe, horrifically, cats were often burned alive with the supposed witches with whom they were said to be on "familiar" terms.

Why all the ambivalence about cats? Perhaps because they are not only highly intelligent, charismatically graceful, and great hunters, but because they seem to be almost supernaturally sensitive. To some this capacity is marvelous; to others it represents a threat, like having someone around over whom you have no control, or someone who, as they used to say in gangster movies, "knows too much."

Actually, it's exactly this supersensitive nature that gives cats the ability to form very special and close friendships with people. With their "cat radar" they pick up on our emotions and mood fluctuations, sometimes even before we are consciously aware of them ourselves.

Like the truest of friends, they are capable of total empathy and loving attention while at the same time never losing or subjugating their individual identity.

My own observation of the extraordinary qualities of cats began before I could even really talk. I had just turned three and my parents and I were living in Canada's Northwest Territories. My father was working for my grandfather's gold-mining operation and we lived in a sixteen-by-twenty-foot tent with a wooden platform for a floor. In the evenings a kerosene lamp threw shadows on the canvas walls, and after supper our tabby would lie calm and sphinxlike, basking in the warmth of the wood stove.

In my memory, the kind of memory of very young days where it's hard to separate what you've been told from what you really do remember, I see an image of her that is like a movie clip. She is crouching in a corner of the tent having a stand-off with a deermouse. I can see the deermouse with its oversized, delicate ears, sitting on its haunches with its front paws up and ready, confronting the cat like a boxing champ.

The cat, whiskers aquiver and tail lashing, turns her green eyes to my mother, who shakes a finger and says in a firm but conversational tone, "No, I told you, outside you do what you want. In here you leave the mice alone."

The cat gets up and goes over to the tent flap to gaze out, acquiescing to my mother's wishes but also, with the eloquence of her turned back, indicating her annoyance.

I remember, too, that same tabby prowling belly low among the grasses, as wild and sleek as a miniature panther. Perhaps it was because of these early impressions that I was later able to see the basic relatedness of all cats, from the many bewhiskered house companions who have lived with me over the years, to the exotic zoo cats large and small that I have been privileged to meet around the world.

EXOTICS

Whenever someone asks me about the nondomestic or "exotic" cats I've known, I always seem to think first of a young serval named Speedy and of three baby snow leopards I worked on in the Zurich Zoo. I think of the snow leopards because of their mythic aura and because of the very special way they recovered from a respiratory infection. Speedy springs to mind so quickly because that's exactly what he was—*speedy*.

When I first saw Speedy I thought he was beautifully elegant. He stood about two feet tall with a slender build and an orangey-brown coat splashed with black spots. His neck was long, his head small, with large rounded ears and pale yellow eyes that watched me carefully. The most striking thing about him, however, was that when I saw him he was rarely still.

Servals are solitary, highly nervous and active creatures. Nocturnal hunters in their African savanna habitat, they are flash sprinters, dashing after game at the same hot-rod speeds as their larger cousin, the cheetah. The stress of captivity can turn their zippy, loner temperament into hyperactivity, which, it seemed to me, was one of Speedy's problems.

Servals and members of the cat family are not the only creatures who experience stress in captivity. That's why, when Art Goodrich of the San Diego Zoo learned about the TTouch and personally experienced what it could do for animals with stress-related problems such as depression and aggression, he asked me to demonstrate the technique.

The occasion was the twentieth annual conference of the American Association of Zookeepers, of which Art was historian. It took place at the San Diego Wild Animal Park in Escondido, an eighteen-hundred-acre area just north of San Diego that has been magnificently landscaped to resemble several areas of North and South Africa and the Asian plains. It was here, as part of my demonstration, that I was to meet Speedy.

Art, a stocky, blond man with a reassuring air of calm competence, is a master zookeeper who has served the San Diego Zoo for over eighteen years. He is not alone in his concern for Speedy and the zoo's other animal inhabitants. Although zoos with intolerable conditions still exist, San Diego is a celebrated leader in a new movement among zoos the world over. Slowly but surely, cramped, bare environments are being replaced with meticulously constructed habitats where our fellow creatures are respected both for themselves and for what they can teach us about our interconnected world.

My demonstration for the zookeepers was scheduled for a Monday, but I arrived on Saturday because I had been asked to work not only on Speedy, but on several animals who had been elected by the park to represent their species in a special educational program. I had arrived early in order to evaluate the animals and do some preliminary work with them before the presentation. I was in seventh heaven.

Speedy had a number of troubles. Part of his job in the animal show was to demonstrate a serval's spectacular ability to leap. To do this, Speedy would make a dramatic bound from one post to another one six feet away, but he had begun to refuse to take the jump and was going somewhat lame in the right leg. X-rays of the leg and shoulder revealed nothing structurally wrong, and the veterinarians, suspecting inflammation, recommended rest.

With several other handlers I watched the serval in his enclosure. He sat in a corner, licking the cage bars. I was told that he had developed this behavior pattern, as well as "marking," or rubbing against anyone who came into his enclosure as a way of claiming them as territory and establishing control over them, after which he would often bite and become aggressive. In the show, he frequently hissed at his trainer and no longer allowed himself to be led on the line, taking a few steps and then stopping, a few more steps and then stopping again.

I thought that stress had probably shut down the discriminatory ability of his mind—he seemed to be reacting as though under constant threat.

His trainer and I went in to him and he immediately began "marking" us. As soon as I tried to touch him he slipped out of my reach, but with no show of aggression, his yellow eyes guarded.

"As long as he's doing all this marking and fidgeting he's not capable of thinking," I said. "He's just reacting."

One of the most helpful aspects of the TTouch is that once the cells are awakened by the circles and new areas of the brain are activated by fresh neural input, an animal becomes capable of mindful action instead of instinctive or habitual response—he is free to act rather than react.

Clearly Speedy was an ideal candidate for this kind of help. The first step was to connect with the serval physically in a way that he would not find threatening. To touch him directly at this point would be too much for him so I opened my bag and took out two of the tools we often use in such cases—a large turkey feather and a short, rubber stick. We put his collar and lead on him and the trainer held him while I stroked his whole body, alternating between the feather and the stick. Using long, slow, rhythmic strokes, I moved across his legs, chest, belly, and back, slowing my breathing and reassuring him in a soothing voice.

We call this calming use of the voice "toning," and we frequently use it in work with upset, frightened, or stressed animals. It's a wonderful aid not only to quiet a nervous animal but also to establish a connection with him or her. Simply keep up a kind of soft litany, like this: "Goooood, you're really gooood, it's aaalll right." You'll be surprised at how much it helps your animal to relax and cooperate.

One reason for this, I believe, is that drawing out and quieting your words automatically slows breathing and movement. When I use the slow exhalations toning requires, I find it always makes me more fully aware of my breathing, which in turn calms and steadies me. And when I'm steady, the animal I'm working with mirrors that steadiness. In other words, the long, drawn-out breaths you must use in order to tone release tension both in yourself and in your animal.

Certainly toning was an integral part of getting Speedy ready to accept the TTouch. Soon I could sense the anxiety in his body begin to recede, like a system of alarm bells shutting off one after the other. Though still ambivalent, he was now willingly standing still for the first time since I had entered the enclosure. We moved to the other side of the cage into the benign warmth of the morning sun, where he decided it was all right to sit down. The trainer held him by the collar while I continued to stroke him, now using only the stick.

Gradually I moved my hands up along the stick until the proximity of my fingers no longer disturbed him. Then I laid the stick aside, and as I began to work on him with the easeful circles of the Lying Leopard (see page 242), the last of his anxiety left him, as though a final barrier had gone down. He flopped onto his side and lay relaxed yet attentive. I moved to his ears with small Raccoon circles (see page 244) and then went on to his neck and shoulders.

I had just begun to work on his right shoulder when suddenly and without warning he reared up and struck at me, his lips drawn back, exposing his teeth in a snarl. His paw, razor claws extended, whizzed by, missing my arm by inches.

Uh-oh, I thought, bad spot, and I immediately went back to making supportive Lying Leopard circles on his flank, giving him plenty of reassurance with my voice. I wasn't too rattled by the incident because, no matter how calm the situation may appear, when working with a potentially dangerous animal a subconscious part of me always remains alert and ready for anything.

Speedy's intense reaction to my TTouch on his shoulder indicated almost certainly inflammation and soreness there. His limp, refusal to jump, and high tension were all more than likely the result of pain.

I worked on his entire body for the next fifteen minutes, carefully avoiding the right shoulder. As I worked I felt his trust flowing back unimpeded through my hands, but the moment I tried to come near the sore place he drew back, tense and hissing. I needed to get him to allow me onto that shoulder, but to do so I would have to calm the defensive-aggressive response the pain was causing in him.

Through working with animals I had learned an amazing thing about how to calm angry and aggressive behavior, not only in animals but in people, too. Neurologically, there seems to be a direct connection between the limbic system (which controls the emotional centers of the brain) and the lips and mouth. So if you use the TTouch on the lips and gums of an animal who is considered aggressive, who bites, or who behaves uncontrollably, you'll normally see a marked shift in attitude. I've witnessed this many times and in all sorts of cases (page 88).

Naturally, people do get worried when they see that I literally mean to stick my hand into an animal's mouth—among all those lethal teeth! But I don't just rush in foolishly. I make no moves until I have observed the animal for a good while and have built up a close, nonverbal, trusting contact through the TTouch.

For instance, with Speedy I moved very respectfully and very slowly, but with definite, firm assurance, working the Raccoon TTouch (page 244) from his neck and ears to his mouth. It's very important with a frightened or aggressive animal not to mirror back his fear, so instead

I concentrated on my breath connection to him, and on letting him know through the calm reassurance of the circles, that I understood his fear and tension, that I was "hearing" him and would not hurt him.

Finally he surrendered his last bit of mistrust and relaxed his head into my hand as if to say, "All right, I'll do it, I'll *believe* you're trying to help." We had reached a place of mutual understanding, of shared confidence—he trusted me. I was about to commit my fingers to his mouth. I lifted his lips away from his teeth and made tiny little circles on his gums with my index and middle fingers. He lay absolutely still, mouth open to my hand, his breathing slow and rhythmic as a dreamer's. (See photo insert.)

After a few minutes I was able to move directly onto the sore shoulder. This time his only reaction was to flinch slightly and turn his eyes to me for reassurance. I went in very gently between the front legs and lifted the chest and scapula from the inside under the elbow. Using great care I worked around the inflamed area with the Lying Leopard TTouch and then ended the session. It had taken half an hour.

As we left the enclosure Speedy was lying quietly in the sun. His gaze was calm and clear, as though he had just woken up from a refreshing nap.

The next morning I worked with Speedy and all the other animals once more and then we were ready for the late-afternoon demonstration. Many of the zookeepers knew Speedy or were acquainted with his problems, and they were astounded. It was the first time, they later informed me, that they had seen the serval perform publicly without hissing. He showed no nervousness on the lead and quietly allowed the two of us to demonstrate the work on his entire body. He even seemed to be enjoying himself.

A year later, when I inquired about him, Speedy was still doing fine. Several of the other problem animals did well, too, and their unique cases appear in some of the following chapters.

At different times in different zoos the TTouch has brought me close to other exotic cats—a tiger, a lynx, an ocelot, a cheetah. Each creature, in the way that animals do, has taught me something special, but the snow leopards of the Zurich Zoo in Switzerland gave me a particularly inspiring lesson in the power of communication.

Snow leopards are rare and shy, and their name evokes the high magic of their home in the Himalayas. I think Peter Matthiessen really caught their mythic quality in his book *The Snow Leopard*:

> By firelight, we talk about the snow leopard. Not only is it rare, so says GS, but it is wary and elusive to a magical degree, and so well camouflaged in the places it chooses to lie that one can stare straight at it from yards away and fail to see it . . .
>
> The typical snow leopard has pale frosty eyes and a coat of pale misty gray, with black rosettes, that are clouded by the depth of the rich fur. . . . An adult rarely weighs more than one hundred pounds or exceeds six feet in length, including the remarkable long tail, thick to the tip . . . It has enormous paws and a short-faced heraldic head like a leopard of myth; it is bold and agile in the hunt, and capable of terrific leaps; and although its usual prey is blue sheep, it occasionally takes livestock, including yaks of several hundred pounds. This means that a man would be fair game as well, although no attack on a human being has ever been reported.
>
> The snow leopard is the most mysterious of the great cats; of its social system, there is nothing known . . .

My friend Ewald Isenbugel, the chief veterinarian of the Zurich Zoo, is passionately interested in conserving the rare animal species of the earth. As Switzerland's director of wildlife research, he attends a great many international conferences on the subject (and once invited me on a journey to Mongolia to find a special breed of hawk—unfortunately, I couldn't go).

Ewald was always very intrigued by my trips to the Soviet Union. Often on my way home from one of these trips I would spend a few days with him and his family, sitting in their blooming garden high above the city, checking on the progress of the baby turtles he was raising for the zoo or listening to the raucous commentary of a fledgling blackbird his children had rescued.

Ewald was especially keen to hear about the Moscow Zoo, most particularly about the snow leopards. The snow leopards in the Zurich

Zoo were so hypersensitive to their environment that it made breeding difficult and raised the fear that even if the female did produce off-spring, she might kill them in her nervousness. In contrast, the snow leopards in the Moscow Zoo were relaxed and had no problems creating and nurturing little ones.

Perhaps, I told him, it was because the Moscow snow leopards felt at home; they were Soviet after all, having come from the Hindu Kush on the Soviet side of the lofty mountain chain that runs along the Soviet-Afghani border. I was joking, but maybe I wasn't so far off the mark after all—the fact was that the snow leopards did feel comfortable, and maybe it was because of the Soviet people themselves.

Most Soviets are ardent animal lovers. Muscovites flock to the zoo every day. I used to see them standing up close to the cages, talking intimately to the animals as though they were family friends, connecting rather than observing. And the animals, nourished by the attention, would come close to their bars to commune with the people rather than pace nervously or ignore their visitors.

We talked further about the problems of the Zurich snow leopards. What I really wanted to do, I said to Ewald, was show him what the TTEAMwork could do for them. My plan, I told him, would be to work on gentling the young ones so they would lose their fear response and be much easier to handle when sick or injured. Then, when they grew up and became mothers, they wouldn't be so stressed.

Reluctantly, Ewald declined. Although he admired the work we had done with horses, he had trouble seeing how it would apply to zoo animals. Besides, he said, the zoo administration did not permit outsiders to work on the animals.

So . . . I had to wait for the right moment. And it came. The Zurich snow leopards finally did become parents. Ewald wrote to tell me the great news; two males and a female, all doing well. But by the time I was once again in Zurich, six months later, all three cubs were at death's door—suffering from a mysterious respiratory ailment that resisted all efforts at cure.

With nothing left to lose, Ewald was able to prevail upon the zoo's administrators to let me try the TTouch on one of the dying cubs. My time would be limited to five minutes, however, because the keeper was afraid that the mother would become angry if I antagonized her baby or handled it too long. As an added precaution, the keeper would hold the cub while I worked him.

Never mind, I thought, five minutes was five minutes.

Ewald and I entered the corridor area behind the indoor enclosure and the keeper went into the enclosure. He brought out one of the young male leopards, holding him in outstretched arms well away from his body. Sick as he was, the young leopard was kicking up quite a fuss, struggling, snarling, and wrestling in the keeper's arms.

No way I can work on him like that, flashed through my mind, and then all thoughts of the rules I was supposed to work under completely left me. I simply reached out, scooped the cub against my chest, and went over to a chair that was standing against the tiled wall. I sat down, holding the two-and-a-half-foot, silky creature across my lap, his long tail draping down to the floor. From that moment on I totally forgot the outside world; nothing existed but the snow leopard and my hands connecting with him.

He stopped struggling almost immediately after the first few circles, and I noticed that he was breathing in a very labored way. With each breath, mucus was forming bloody bubbles at his nose. I began to breathe in unison with him, making a soft, comforting "shhhhhing" tone with every exhalation. Once I had entered his rhythms and made contact, I began work on his ears, and then as his quietness deepened, I used tiny Raccoon circles all around his sinuses and around his eyes.

After what must have been approximately three minutes, the cub's breathing calmed and the mucus bubbles ceased to appear. He lay still as sleep, eyes half closed.

I continued the circles down his neck, then onto his head, following the outside path of the nasal passages and lifting the lips to work on the gums. I formulated no plan but simply followed the lead of my hands, which seemed to know where they wanted to go and what they wanted to do, transmitters through which poured a silent flow of cellular information.

Moving down, I worked each leg with small Python Lifts (see page 254) and then circled into the spaces between the pads of his paws, down the entire spine, and then out to the tip of his tail. As I worked, the young leopard felt heavier and heavier as he let go into ever-deepening levels of relaxation. His head rested trustfully on my arm.

When I reached the tip of the tail I stopped, sat back, and looked up. To my surprise, I found myself back in the world again. There stood Ewald and the zookeeper, exactly as I had left them. They had scarcely moved a muscle. Later Ewald told me that they, too, had stood there in a timeless place, mesmerized by what was happening, astonished that the leopard would actually permit me to move into that most vulnerable of all places, the delicate area between the paw pads.

Apparently, in what is known as "real time," twenty-five minutes had elapsed. We returned the by now sleepy leopard to his mother, who wasn't in the least bit perturbed and who greeted him with licks and nuzzling.

I had to leave Zurich the next day, but shortly thereafter I received the happy news that the cub had started on the road to recovery almost immediately after his treatment. Not only that, but his two littermates, who had been equally death bound, were recovering beautifully if inexplicably right along with him.

I smiled to myself. I had a secret, but I didn't want to tell anyone because I was afraid they would think I was crazy. The whole time I was working on that young cat I had also been visualizing a cellular connection to his siblings, as though they too were receiving the TTouch. And afterward, when we had returned the cub to his mother, I had noticed that the other cubs had also become calm and tranquil.

Again and again I have been deeply impressed by examples of the power of visualization (pages 138 and 183) and the untapped resources we all have within us.

FELIS CATUS A.K.A. PET CATS

Recently, Ewald sent me a wonderful picture of the three young snow leopards, now several years old, eagerly crowding up against their wire mesh fence to receive their shots from him (see photo insert). (He finds the TTouch has made them easier to handle and more amiable.)

Their beautiful, intelligent faces remind me of spectacularly exotic pet cats like Abyssinians or Balinese. Because wild and domestic felines, both great and small, do have such strong family resemblances, people sometimes get confused and try to make pets out of jaguars, ocelots, or cheetahs, a dangerous and misguided idea.

Not only are these breeds inappropriate as pets, it is illegal to own them and many of them are sold by smugglers who ship them in such careless and cruel ways that they often die en route. It's much safer—and also more humane—to stick to *Felis catus*.

Those who don't like felines often say that "cats don't care about you." Millions of bedazzled cat owners, veterinarians, and psychologists will attest to the fact that this is simply not true. In my own parade of cats there have been many whose concern and sympathy for me bordered on the maternal.

Cats, however, *are* an independent breed and some are inclined to remain on distant though not unfriendly terms with humans. So, what do you do if you chose a kitten who turns out to be one of these—when you wanted a cuddler?

"THE UNTOUCHABLES"—CATS WHO ARE INDIFFERENT OR AFRAID OF US

The first cat I tried the TTouch on was a stray. He was a tiger tom with wonderful, big jowls, extra large paws, and a husky, strong body. I just loved him. He, on the other hand, cared only about breakfast.

I was living in the famous California resort town of Carmel, a place that always reminds me of villages in fairy tales; the houses look like Hansel and Gretel cottages and tiny shops that serve tea and crumpets dot the streets. At the base of the six-block main street you can see the ocean, and at night the trees on this "boulevard" are lit with hundreds of little lights, like fireflies.

The residential streets have no streetlights; they're dark at night and very quiet. I lived two blocks off the main street and from my house I could take a dirt road down to the beach and the ocean, or walk for miles on the many paths that crisscrossed the nearby forest.

Every morning, at seven o'clock sharp, the big tiger tom would appear at my door and ask to be let into the kitchen for his breakfast. He'd march in, eat, and leave, very aloof, not interested in being petted and certainly not about to allow himself to be picked up.

I was away often then, sometimes for months, traveling around the world teaching, yet on the very day of my return, there he'd be at his post by the kitchen door, meowing for his breakfast as though he had received a telegram announcing that I was back. And then we'd start our old routine all over again.

One day as he was standing by the door licking his breakfast delicately off his whiskers and waiting to be let out, I thought to myself, This is really crazy. Why should I feed this cat all the time and not at least be able to hold him?

So I closed all the doors to the kitchen, and caught him—not an easy thing to do I might add. I spent quite a few minutes following him around the room over and under chairs until I finally got hold of him. I restrained him on the table, not tightly enough for him to panic, but enough to keep him there for all of three minutes. He didn't scratch or bite me, he just glared and struggled to get away.

I managed to get in a few fast circles all over his body. He didn't like it at all and stalked out the door looking ruffled, but he must have thought it over, because he showed up again the next morning and this time, when I repeated the procedure, he was much easier to restrain.

On the third day he sat quite still for large, very light Lying Leopard circles, closing his eyes until suddenly I was mindful of a different kind of awareness. A quiet stream of nonverbal communication had begun to flow between us. It was very subtle, a heightened perception a little like that first moment of acuity when your ears pop open after pressure in an airplane.

Afterward he hung around for several hours.

On the fourth day he not only stayed for breakfast, he stayed for the rest of his life.

When faced with a nervous cat—or any nervous creature for that matter—the best way to introduce the TTouch is by jumping all over the body with fast circles so that the animal doesn't know where your hand will land next. This focuses the animal's attention on you at the same time as it breaks the pattern of nervous reaction. And finally, because the animal can't anticipate, he or she very quickly gives up struggling.

If a creature is really very upset and nervous at the start, don't make your circle completely closed—make several incomplete circles, beginning at six o'clock and ending at four. As soon as you feel your touch being accepted a little more easily, move on to the basic closed circle.

The wonderful thing about Tigger was how friendly he became. Not only did he sleep with me, but later, when I moved to a ranch in Naciemento, California, where I was running a retreat center and residential school for horsemanship, he would make the rounds of the guest rooms and give each guest a visitation and a cuddle. He was a truly democratic cat and never left anybody out.

After he became a close friend he used his feline psychic abilities for more than simply knowing when to show up for breakfast, as became apparent when I was making the move from Carmel to Naciemento. At first I had doubts about taking Tigger away from his home community and his daily round of backyards and forest hunting grounds. Perhaps it would be better to leave him with a friend in the neighborhood.

I made three trips to Naciemento with the moving truck. Tigger was nowhere to be seen. He's hiding because he doesn't want to come along, I thought sadly.

Finally, the moment came for me to lock up and leave my storybook house with the Dutch doors and the blooming garden. I went out to my car feeling bad about Tigger, who had disappeared shortly after breakfast. But as I stepped around to open the door on the driver's side, there he was, sitting pertly beside the front tire. As soon as I opened the car door he brushed past me, hopped in, and looked at me as if to say, "Well, what are we waiting for? Let's go." He knew before I did that the time had come for him to declare his loyalty.

Tigger had begun by not wanting to be touched. Some people find that haughty aloofness attractive, but what most of us want from a cat is a cozy relationship full of warmth and affection.

Perhaps you have a cat that tolerates you but only for short periods; you put the cat in your lap and two minutes later she wants to get down. Or your cat is a one-person cat who won't allow anyone but you to pick him up, not even your husband, your kids, or your friends. My coauthor, Sybil Taylor, had such a cat, a big, bearish gray named Pooh who used to have eyes only for her. He would drape himself around her neck like a furry scarf, but if anyone else came into the room, let alone tried to cuddle up to him, he would streak for the depths of the nearest closet as if pursued by the hounds of hell.

When I first met him, I picked him up and told him I was going to introduce him to the pleasures of becoming a social cat. "Believe me,"

I told him, "you'll be glad we did this." But as I looked into his alarmed yellow eyes I could tell he was anything but convinced.

At first, though he glanced longingly at the closet, he permitted me to lift him onto my lap and hold him briefly, while I used my arms and hands as boundaries for a light restraint. In seconds, however, he began struggling to get down, so I decided on another approach.

Taking a towel, I wrapped him in it, using one hand to hold two edges together over the back of his neck like a collar, and tucking the rest of the towel around his hind legs to contain him. With my other hand I went over his body with randomly placed Raccoon TTouch circles. After he saw that his struggles did him no good, he quieted down, and in a few moments I was able to remove the towel and hold him with one hand supporting him under his chin and the other hand doing the TTouch (see page 59).

A week or so later, after daily ten-minute sessions, I entered Sybil's workroom to the amazing sight of Pooh stretched languorously on the table, offering me his belly for a good tummy rub. Now, a year later, while the cat is still sometimes shy with strangers, he has become so social that he joins in at after-dinner conversations, watching and listening with obvious pleasure from his own chair at the table. After a while, he gets up, sidles over to the nearest person, and with a mellow glance, sends out an invitation to be picked up.

This technique is very helpful in preparing a cat to enter a carrying case without a battle, to travel in a car, or to visit the vet. The TTouch is very effective, too, as a way of reducing the stress and fear that often accompany presentation at professional shows. It is also beneficial for cats who visit hospitals and rest homes as part of "animal therapy" programs, as it prepares them to relax and enjoy the attentions of the many strangers to whom they bring pleasure.

BEFRIENDING FERAL CATS

Greenacre Farms in Santa Cruz, California, is a small Arabian-horse breeding farm. I had been invited there for a "clinic," a training work-shop that lasts several days. People often bring along their problem horses and work with our TTEAM methods.

We were a group of ten, a diverse bunch including a psychologist, a dancer, several trainers, a school teacher, an acupuncturist, and a writer. I like to give my talks outside whenever possible, with everyone sitting on the ground in a circle, the most natural shape for warm and harmonious communication. It was a sunny day and the spice-jar smells of earth and trees wafted deliciously around us. A little distance away, two kittens bounced around in the grass, stalking each other.

I've noticed that many animals are drawn like magnets to groups of humans who are meditating or concentrating quietly together. I've seen it happen over and over again in workshops all over the world; almost inevitably a dog or cat will come trotting over and sit next to someone or flop down happily right smack in the center of the circle.

But here at Greenacres things were a little different. The two young gray cats had been attracted to us, yes, but instead of joining us, they stayed cautiously outside our circle. Ten months old, they were part of a skittery litter that was growing up in the stable, catching mice, playing hide-and-seek in the hay, and dodging the flashing hooves of the huge creatures all around them.

The only humans who could approach within six feet of these adolescent wild ones were our hosts' children, Abigail, two years old, who followed the cats around with the determined waddle-waddle of a kid in diapers, and her serious-faced older brother, four-year-old Peter. But even though the children could hunker down beside the kittens and tease them with long stems of grass, at the first sign of a hand reaching out for a pat, the creatures were gone.

To me, teaching in a clinic is more than simply sharing what I've learned from all the animals I've worked with in the past. The exciting thing is being able to use what I've learned spontaneously so that I'm on a voyage of discovery right along with the people I'm teaching and the animals we're learning from. And so it was that the two kittens rolling around like furry wrestlers inspired me with a new idea.

"I'd like to show you how I work on young foals or horses who have never been touched," I told the group. "The principles are pretty much the same with foals as they are with kittens or with any animals, except that with kittens you can take advantage of the fact that they'll play with practically anything that moves.

"Here are your tools," I said, holding up two of our wands (see page 270) and walking over to the kittens. Of course they instantly bounded away to sit down and stare at me from the edge of their six-foot safety perimeter.

I stretched out an arm and, with the added length of the wand, I was able to reach across to the cats without violating their demilitarized zone. I dragged the wand on the ground about three feet away from them and one of the kittens, the feistier one with a slightly bent tail, immediately pounced on it.

Then I took the other wand and playfully teased him into chasing both wands. After a while I was able to stroke him briefly and lightly on the back and shoulders. He was frightened at first, but within three or four minutes I could stroke him all over with one wand while he was busy playing with the other. Then I was able to take the two wands and stroke him with one on either side.

He was completely intrigued by this sensation and I took advantage of the moment to turn the wand around to the thicker end, with which I then began to make small circles on his shoulders.

During the next ten minutes, I continued to move my hands down the shaft of the wand until I was able to perform a few fast circles directly on his wiry little body. Finally, after I had worked down the wand a second time, he was entirely open to the TTouch.

We printed the story of these wild barn kittens with photos in our next TTEAM newsletter (see page 18), and subsequently received numerous case histories from people who had read it and applied the technique successfully in all sorts of situations dealing with feral cats. We heard, too, from a number of people who work in animal shelters who told us that this double-wanded taming technique is extremely helpful. Dr. Tom Beckett, veterinarian for the Austin, Texas, animal shelter, reports that he has used this TTEAM approach on literally hundreds of cats brought into the shelter. In fact, he has found it so useful that he intends to deliver a paper on the use of TTouch for feral cats to the Delta Society, an international organization dedicated to the research and promotion of the human-animal bond.

Dr. Beckett's partner, Marnie Reeder, told me that they've had a great deal of fun at the shelter taming wild kittens. The kittens are kept in the bathroom and two wands are left by the door, so that any time anyone goes in, he or she spends a few moments stroking the kittens. The kittens are also stroked at feeding time. After five to ten days the formerly wild ones are not only tame, they have discovered the pleasures of being handled. They are now suitable for adoption and, as Marnie puts it, "ready to become successful members of society."

AGGRESSION AND PLAYFULNESS

One of the ways that kittens and puppies learn about their world is through play with their littermates, a rough-and-tumble proposition. People often think that when they play with their animals they have to imitate the animal, that they have to become *like* a kitten or puppy, a cat or dog. What they're actually doing is making their pets difficult to live with, because the message an animal receives from unruly play is that it's all right to bite and scratch and bark. It may seem like innocent fun to roughhouse with your pup or kitten, but what is cute in a baby can become unmanageable behavior in a mature animal.

I'm not saying you can't have a wild and wonderful time with your kitten; just don't offer your arm as a clawing and kicking post for her back legs, or your hand as a teething ring. Play with her gently, and for those wilder moments of abandoned delight use a ball, a length of string, some crumpled-up paper, or a dangling pencil—whatever you dream up, as long as it's not a part of your body.

There's a thin line between playfulness and what can appear to be aggression, and sometimes, when an animal seems to be playing too fiercely, she may actually be telling you that something is physically or emotionally wrong with her. I learned a lot about this from a baby ocelot in the Los Angeles Zoo. Zelda was four months old and had been weaned early so that she could be trained for the zoo's wild animal demonstration shows. The trouble was Zelda was unteachable; all she wanted to do was play.

She didn't mean to be aggressive or injure anyone. She just became very excited and overactive with her mouth and feet whenever Holly, her keeper, tried to handle or teach her.

The first time I went into her enclosure I had the same problem. I sat down on the ground with Zelda and contained her lightly with my arms, not holding her, just creating a limited space. That was all right, but as soon as I tried to touch her she was all over me, mouthing my arm, my hand, dying to play. It didn't hurt—Zelda was just a roly-poly ocelot kitten weighing about twenty pounds. Her teeth were baby teeth, but still it was frustrating for both of us.

I tried a number of different positions, but nothing worked; no matter how I approached her she continued to play, kneading me frantically with her paws. Holly shook her head and I could feel myself beginning to sweat and grow upset.

You can be the most loving, easygoing trainer in the world, but when, over a prolonged period, you can't get an animal to cooperate, you can lose your perspective and start thinking that *you're* solely at fault. You start believing that what's happening is a reflection of your inability to get the job done, and so you feel threatened and frustrated.

When that happens, a certain tension builds up. To release it, I sat back and took a few deep breaths. Breathing centers you. As soon as I recognized and let go of my own need for control, I suddenly had the impression that Zelda was connecting to me. All at once, as though a fog had lifted, I got a clear picture of what was bothering her.

What she wanted with her playing was *contact*. She was desperate for it. Zelda had been taken away from her mother when she was very young. She had nothing soft or cuddly in her enclosure, no warm mother's body to knead with her paws and nurse with her mouth, no other young animal to push and pull and play with. What Zelda was doing was expressing her social deprivation. It was the only way she knew to connect.

So instead of letting her play with me, I took deliberate hold of her legs and feet, making conscious, solid contact, and slowly rotated each small leg, making deep circles between the pads of her paws. I went into her mouth and pressed hard on her gums and pushed against the roof of her mouth, letting my intuition effortlessly guide the motion of my hand.

After fifteen minutes she had calmed down. She was receiving what she had been signaling for so desperately. When our animals resist us or don't cooperate with us we just have to look to figure out what it is they need, what it is we can do for them. But of course in doing so we also have to be careful to set boundaries.

Sometimes, with animals that are aggressive, nervous, or overreactive, you have to find something they like and use it as a kind of enticement. For instance, your pet may accept work on the ears but hate to be touched on the back. In that case, you should slide your fingers over the ears with one hand while the other hand is working quietly doing Belly Lifts, or very calmly and slowly doing the Lick of the Cow's Tongue (page 250) under the belly up to the spine.

With this technique, animals begin to associate something they like with something they can't stand, and so slowly learn to accept the contact.

Using the TTouch on very young animals lets you make the desired strong physical connection with their bodies without going too far. Your physical games with them can become gentler, but just as much fun, because you have given them a way to relate to you while at the same time satisfying their natural need for physical contact.

AGGRESSION AND TERRITORY

Just as the instinct for play can border on aggression, so territoriality or maternal protectiveness can trigger attack. I had experienced this phenomenon in cats before, but it took Sveata, one of the most unforgettable feline characters I've ever known, to really teach me about it.

I first saw Sveata at the Children's Museum in Santa Fe, where I was teaching the TTouch to a group of preschoolers. The animals we were working on had been sent to us by the Santa Fe animal shelter and Sveata was one of them. The shelter's director had chosen to send her because she felt sorry for her; that morning it had been necessary to euthanize Sveata's newborn kittens and she was clearly very upset. Hardly more than a kitten herself, she weighed less than five pounds and her small body was so narrow it looked as though an Egyptian profile painting had been lifted off the wall. She was gray with a white nose, a white chest and boots, and an abbreviated three-inch tail that stood straight up in the air.

Sveata wasn't her name when I met her—I called her that later because it's a Russian name and she reminded me so much of another gray cat who had befriended me in Moscow. At the end of the teaching session I discovered I just couldn't bring myself to send Sveata back to the shelter, so I took her home, where it quickly became obvious that I had been chosen to replace her lost kittens.

We would be sitting on my couch, French doors open to the summer breeze, Sveata stretched out beside me like a royal temple cat. Suddenly one of the community cats or dogs would wander by outside and off she would be, out the door and on the attack. No other pet was allowed to approach within fifteen feet of my house.

There are a number of community dogs and cats who have the run of the property, so Sveata, my self-appointed guardian, was going to have to learn to change her ways. Immediately after each "episode," I would bring her back, work her ears and body, and lightly restrain her in my arms. After a few sessions she began to lose the fear that I was sure lay beneath her desire to protect me.

One night, a few weeks after Sveata came to live with me, I was awakened by loud crunching from the direction of the French doors, where I kept her dish of food. Strong jaws for such a little cat, I thought, but then I saw Sveata's head pop up alertly from her position at the foot of my bed. Uh-oh, I thought, trouble ahead; "whatever" was doing the crunching would have to be dealt with very quickly and very quietly.

"It's O.K., Sveata," I whispered. "I'll take care of this." Very slowly then, I sat up, and in the moonlight I could just make out a chunky black-and-white figure and a plumed tail swishing backward and forward—our midnight snacker—a skunk.

Normally, my attack cat would have been in full action by now, but though she looked crouched and ready to pounce, she remained motionless on the bed while I tiptoed over to the skunk. "Now, you're just going to have to go," I told the skunk, making slow sweeping motions behind it with one hand and gingerly beginning to close the door after it with the other. The skunk allowed itself to be gently propelled halfway out, but as it reached the middle of the doorway it decided to pause, perhaps to think matters over. I just kept right on slowly closing the door and praying there would be no big burst of noxious spray to linger forever in the porous adobe walls and brick floors of my room.

Finally our uninvited guest was gone. I looked at Sveata; though she was still ready and waiting, she had not moved a whisker.

Such behavior, however, was unusual. Although Sveata now permitted other creatures to move unchallenged outside our house, she still would not allow another cat or dog to cross my threshold without a major ruckus. This made her a perfect candidate for a training workshop I was scheduled to lead on the TTouch for companion animals.

Companion animals are those that share our lives but do not work for us. Half the people who registered for the four-and-a-half-day workshop had dogs and cats to whom they wanted to relate more effectively. Some of those attending had animals that were growing old, some had small creatures like rabbits and gerbils about whom they wanted to know more, and the rest were TTEAM practitioners for horses who wanted accreditation to teach the TTouch for companion animals.

We held the workshop on the second floor of a nineteenth-century adobe mansion, a large old jewel of a house that was scheduled for remodeling. The huge upstairs room was twenty by fifty feet with a ceiling of enormous hand-hewn log beams. We turned the empty space into a bright work place, bringing in rugs and flowers. We had asked the Santa Fe animal shelter to bring to us animals with difficulties that made them unadoptable, problems like fear biting, incessant barking, unusual nervousness, and unpredictable aggression. I brought Sveata.

We worked on several animals every day. When it was Sveata's turn and she saw the other cats in the room she instantly began to hiss. One of the participants had brought her own cat, a very sweet-tempered black female, so we sat down on the rug with the two cats a few feet apart and began to work on them simultaneously. I put a big, green stuffed dragon between the cats to block their view of each other.

Every time Sveata growled, I worked her ears and got her to lower her head. When she stopped growling we narrowed the space between the two cats by a few inches, moving the dragon in closer as well. The dragon was a marvelous felt puppet with a red tongue and a head into which you could insert your hand to move the mouth and nose. I put my hand inside the head and began to make circles on Sveata with the dragon's nose, reaching over after a moment to do the same to the black cat, and then back to Sveata again.

We continued this, moving the cats closer and closer together. At the end of twenty minutes both cats were less than eight inches apart, totally relaxed and enjoying themselves. I let Sveata get up and walk around while I kept contact with her by stroking her back lightly with a wand. She was tranquil and made no aggressive move at all toward the other cat. I then passed her around to be worked on by several different people in the group.

The following week Sveata was much more accepting of the other animals that came to visit, and whenever she did growl, I just picked her up and worked on her ears. Another week passed and I came home one day to find no Sveata. She was next door visiting the neighbor cat and sharing a snack out of his dish. He was an easygoing guy, so he didn't mind. Several days later, when he came over to my house, she returned the favor.

If you are having trouble with your possessive or territorial cat, try following the steps I used for Sveata, and if you don't happen to have a green dragon puppet on hand, improvise.

THE TTOUCH FOR INJURIES

▷ *Shock*

Often, when an animal or human is severely injured, the body goes into shock, a dangerous and sometimes fatal condition in itself. The pulse becomes elevated and the rest of the body processes slow way down. Miraculously enough, there is a simple way to bring the body out of this precarious state. The secret, known to Chinese acupuncturists for centuries, lies in the ears. According to acupuncture theory (which is increasingly accepted by the Western medical establishment), the ear (and the foot, too) is a miniature representation of the whole body. On the ear are "points," or places where life energy can be strongly activated, with each of these points corresponding to a different part of the body. Thus if you stimulate the whole ear you are, by correspondence, stimulating the whole body. There are also particular points on the ear that, when stimulated, can help counter specific physical disorders.

The TTouch utilizes both of these concepts. We found that when you work the entire area of both an animal's ears you greatly enhance the functioning of all the systems of the body, giving it a powerful boost. We also learned that certain locations on the ears, when worked with the TTouch, produce very specific and helpful results.

The point for shock is located at the top tip-point of the ear for horses and most animals. When an animal (or human) goes into shock, you can make an impressive and often lifesaving difference by instantly working the ears, paying particular attention to the tip.

The matriarchal, aged gorilla at the Zurich Zoo who first inspired Linda to work with zoo animals (see page 9).

Roger Russel

Linda (third from left) and her family, "The Riding Hoods," in 1953 at the Edmonton Spring Horse Show in Alberta, Canada. Sister Robyn is the baby on the far right being led.

Yellow Knife, Canadian Northwest Territories, 1938: Linda with her aunt Helen and a bear cub at her grandfather's gold camp. Notice the fingers on the cub's ear—intimations of things to come.

Linda Tellington-Jones, 1988.

Stevi Johnson

Linda working in the mouth of Speedy, the African serval (see page 33). Speedy had a habit of licking the bars of his cage and hissing when confronted with crowds of people. Working the mouth changes patterns of emotional behavior.

Stevi Johnson

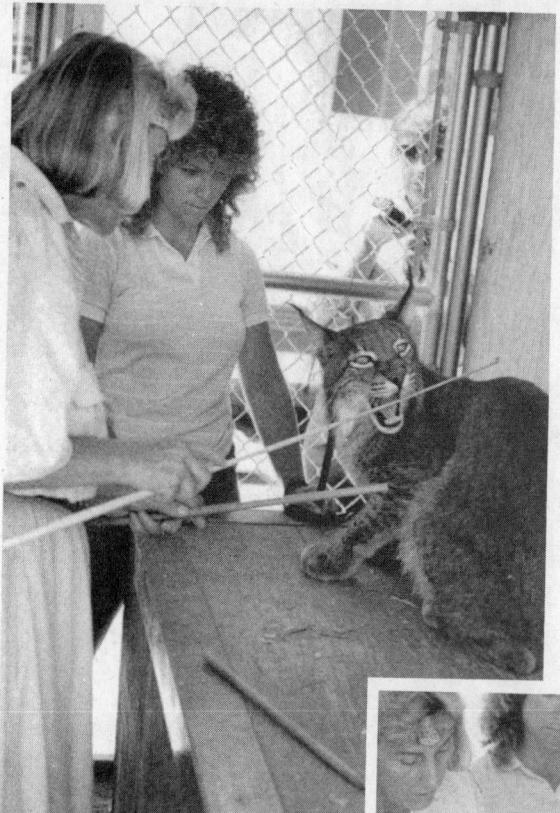

Kashan, a lynx, had begun to show signs of aggression and lack of cooperation with his trainer. Here, Linda begins the TTouch session with two wands to create distraction and to allow an approach that is safe for her and nonthreatening to Kashan.

Stevi Johnson

Twenty minutes later, Kashan relaxes with the TTouch on his ears.

Stevi Johnson

Bud is an Animal Ambassador who teaches interspecies communication to the children of the Meridian Elementary School in Boise, Idaho (see page 74).

Ann Finley

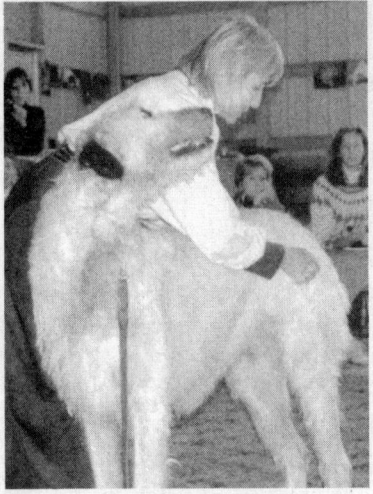

Carol Lang

Working on the show dog Prince, an Irish wolfhound. His performance lacked confidence, more than likely because he had bone cancer. Owner-trainer Terri Raeder reported that his jumping improved and he was much more comfortable after Linda worked TTEAM and TTouch with him.

Linda does the TTouch on Candy, a hyperactive cocker spaniel puppy, to calm and focus her.

Stevi Johnson

"Be careful, Mindy—let's work this out": Linda begins a TTouch session with Mindy coyote (see page 65). The dowel stick is used as a first step in reaching out and stroking her to lessen her feeling threatened.

Mindy attempts to prevent Linda from making contact with her. Two short rubber hoses allow Linda to work closely with an aggressive animal, while at the same time maintaining her own safety. After less than an hour's work, Mindy willingly presented her flank to Linda for the TTouch.

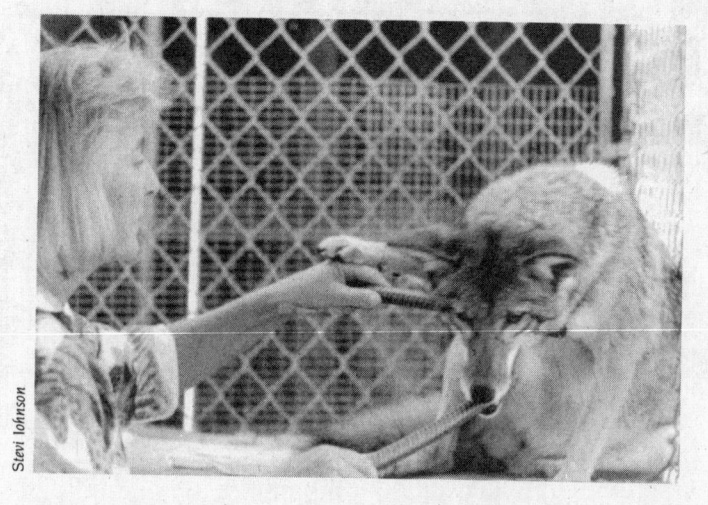

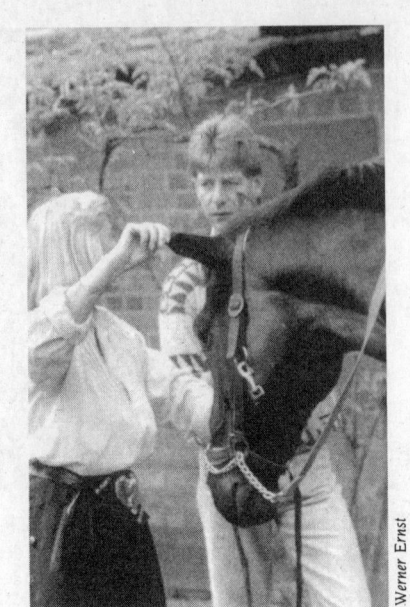

Linda, German Olympic Gold Medal winner Claus Erhorn, and his horse, Justyin Thyme. Linda is working the horse's ears, demonstrating a method for rejuvenating horses after hard work.

Werner Ernst

Valerie Santagto

Linda and Claus Erhorn working with the TTEAM exercise called the Homing Pigeon (see page 272). In this case, the exercise is being used to calm the high-strung mare and to lengthen her stride.

Denise Lynch

Linda bareback on Denise
Lynch's Arabian gelding, Harah.
Riding with no restraint on
the horse's head deepens
communication and joy
between horse and rider.

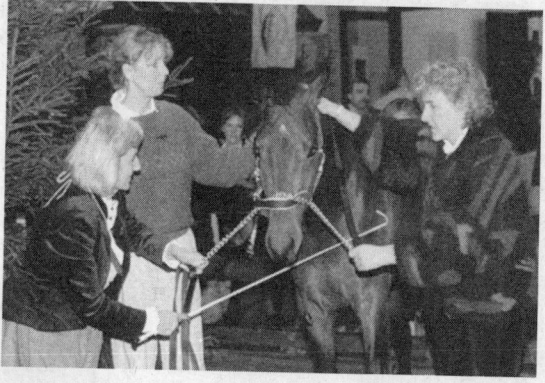

Werner Ernst

Linda (left) and her sister Robyn Hood
(right) demonstrate the TTEAM work at
Equitana, the famous German equine
trade fair. European TTEAM Director
Annagret Ast assists. Note the TTEAM
chain and halter. Leads over the nose
and at both sides keep the horse
calmly attentive.

Werner Ernst

Communication!

Linda slowly strokes Archimedes Owl (see page 132) at the San Diego Wild Animal Park Critter show. Using two feathers focuses and relaxes birds, and prepares them to allow and enjoy the human contact of the TTouch.

Stevi Johnson

Stevi Johnson

Working with the Raccoon TTouch to encourage
the regrowth of feathers on the legs of a cockatoo.

Case in point: my brother Gerry was at work one afternoon at our family's peat moss supply business when a phone call from his ten-year-old daughter, Karen, sent him racing to his car. Frantic and sobbing, she begged him to come home—their cat, Blue, was pinned under the automatic garage door and was dying.

Blue, a five-year-old Russian blue, and the darling of all four of Gerry's children, had been at the wrong place at the wrong time: no one saw him wander under the door at the exact moment that one of the kids pushed the electronic button to close it.

As Gerry ran up the driveway he saw the cat's head sticking out from under the bottom of the door. A knot of kids and neighbors stood around helplessly and from inside the house came the sound of a child's hysterical tears.

"It was terrible. I'll never forget it," Gerry told me later. "Blue's eyes were glazed over. I thought the cat was dead. But then I heard a moaning sound, so I grabbed an iron bar and pried the door up a fraction, and Blue leaped up and tried to run away. But all he could do was go around in wonky circles before he finally fell over."

Gerry grabbed up the cat and put him in the front seat and drove to the vet. On the way, poor Blue kept struggling to sit up, but with each try he'd promptly fall over again. Gerry felt completely helpless, afraid the cat was going to die right there on the seat beside him. Suddenly he remembered that I'd said working the ears was good for shock, so steering with one hand, he worked Blue's ears and neck with the other, keeping up a steady stream of encouraging talk. The cat immediately quieted down and became calmer and more comfortable, which of course helped Gerry's anxiety, too.

Blue didn't die. He had a concussion, from which he recovered perfectly several days later. He's completely normal now—except for the fact that he avoids the garage.

▷ *Rehabilitation*

When the ten-month-old white kitten was brought into the emergency room of the small animal hospital where Christine Schwartz works as an assistant, she thought the cat had come to the end of all her nine lives. She had a broken jaw, a severely smashed face, both her eyes were swollen shut, her left ear had been torn off, and the left side of her brain was swollen and pressing against her skull.

Her name was Mitten and she had been mangled by the fan belt of a car engine in a bizarre accident that, Christine said sadly, happens all too often. Cats in cold climates like warmth and will frequently climb up into the engine of a parked car to hide away under the hood for a nap. The owner starts the car without realizing the cat is there and the animal winds up dead or in the trauma room of the veterinary hospital.

Mitten remained in shock and unconscious for two days. The swollen brain and abnormal constant purring strongly suggested severe brain damage. When she did awake, Mitten was blind. She couldn't eat, drink, swallow, or urinate, and she had to be restrained from turning constant circles to the left in a pathetic attempt to get away from the pressure inside her skull.

Christine had worked on our family farm for years and had taken many TTEAM training courses. She did the TTouch on Mitten's ears and body as soon as Mitten was admitted, despite her fear that the cat was too badly injured for it to do any good, and yet she continued administering the TTouch every three hours, thinking that it might at least make the cat more comfortable.

After two days everyone at the clinic was surprised at the kitten's dramatic and ongoing improvement. Her constant purring stopped whenever Christine worked on her, and after ten minutes of the Raccoon TTouch (see page 244) on her mouth and throat, she was able to lick baby food and swallow. Delicate work around her closed eyes caused her to open them and respond to the people around her.

After yet another forty-eight hours, the inappropriate purring was much reduced, and by the third day it was gone entirely. Mitten had also stopped her obsessive circling and was actually able to limp around the room. Her vision, however, was still a mystery to Christine. While the kitten did turn her face toward people, she seemed unable to focus her eyes. Her coordination was very poor; she walked around the room bumping into things and often became confused about which foot to move when.

Christine then took sticks and rulers and built a miniature labyrinth and star (see page 268) to retrain the cat's neural pathways. After a week of daily ten-minute sessions in the labyrinth, Mitten had a much better sense of where to place her feet, and her overall coordination had improved markedly. In fact, Christine reported to me, this kitten, who had almost been given up as hopeless, was eventually able to go back home.

THE TTOUCH FOR OLDER CATS

I didn't know Angela very well; she had asked me to lunch after a TTEAM teaching clinic, but as soon as I saw her house, I felt good. It looked welcoming, a white colonial building embraced by fragrant evergreens on both sides, with a bright-leafed copper beech and two slim young birches on the front lawn.

Inside, the atmosphere was soft and restful, the color scheme all muted pastels and earth tones. During lunch a white Persian cat strolled through the room, sniffed alertly at my riding pants for news of the horses, brushed purring against my hostess, and strolled out again. A few minutes later I glanced out the window to see a flash of white tearing through a bed of orange fire lilies.

"How old is your cat?" I asked. "About two or three?"

"Gracie is eighteen," Angela said, with an amused glint in her eye.

I was shocked. "That's amazing. I've never seen an eighteen-year-old that healthy. But you know, I'd like to suggest that you work on her ears, because if you'd do that, she'd probably go on forever."

"Oh, it's funny that you should say that." Angela looked even more amused. "Almost every day of her life I've just sort of intuitively worked the inside and outside of her ears between my fingers."

Many people are intuitively attracted to working on their animal's ears because "it just feels so right." It delights them to find out that they were actually onto something and that with just that little difference in touch and attention they may be helping to create a healthier and longer-lived pet.

▷ Wimsey

Wimsey, an eleven-year-old cat belonging to TTEAM practitioner Barbara Janelle, was very good at vocally expressing her opinion. She was half Siamese and half barn cat, and as cat lovers know, Siamese are big "talkers." She was also, Barbara says, an exceptionally wise and kind old cat.

Wimsey used to stay out in the backyard all day doing the things that cats like to do, and then come in for the night, but as she got older and stiffer she began to stay in more and more, preferring the tamer comforts of couch and carpet. Barbara noticed, too, that she was having trouble climbing stairs and was no longer as willing to leap up on beds, chairs, and shelves. Her back above the haunches had developed an arch and the fur along the spine stood up in spikes.

Barbara began doing the TTouch body work on Wimsey, and to her amazement, after a few sessions the cat would come over, stretch out, and literally present the parts of her body that needed work. "I found the Tarantula particularly powerful," Barbara told me. "I've worked with many cats and it always seems to be especially effective."

After several sessions Wimsey regained much of her agility and fluidity of movement, even galloping around the house with Barbara's other cat, a frisky two-year-old, and leaping to a favorite perch on the kitchen shelf. Barbara instituted regular TTouch sessions and was delighted to see Wimsey's rumpled fur sleek and smooth once more and the arch in her spine much improved.

If you have an aging cat, try beginning with the Lying Leopard TTouch (page 242) to create a feeling of support and warmth. As you move upward to either side of the cat's spine, it's important to keep your pressure very light as the vertebrae of cats are particularly sensitive. In fact, while you can work close in to the vertebrae, it's best to stay off the spine itself entirely.

As Barbara recommends, the Tarantulas Pulling the Plow (see page 256) is a wonderful TTouch for older cats. Cats have loose skin and the rolling sensation is both pleasant and stimulating. Also effective is the Clouded Leopard, a supportive TTouch again done by giving the animal contact with both your fingers and palm (see page 241). Gentle Belly Lifts (page 258) are also helpful for a cat who no longer gets much exercise, or who has a poor appetite, followed by the Snail's Pace (page 245) upward toward the spine.

Of course, no one knows your cat as well as you do, so remember to listen with your hands; your cat will help you find just the right movements to make.

TAKING TIME TO TTOUCH

Instead of stroking your pet while you're watching TV or talking to someone, take five minutes a day to do light Belly Lifts and TTouches and work the ears.

Stroking can quickly turn into an unconscious, repetitive act, one that can become boring to you and even irritating to your cat. When you hold your pet in your lap and do the circles, however, you're not only making your cat feel good, but because the circles focus you, you're setting up a deeper level of communication than is normally possible.

Using the TTouch all over the animal's body will not only help the cells to function more effectively, it will also aid your cat's entire nervous system in maintaining health and well-being.

Now that we have explored a number of ways of using the TTouch with cats, I hope that you'll find yourself using it more and more frequently—and not just with your own special friends, but with all cats that cross your path.

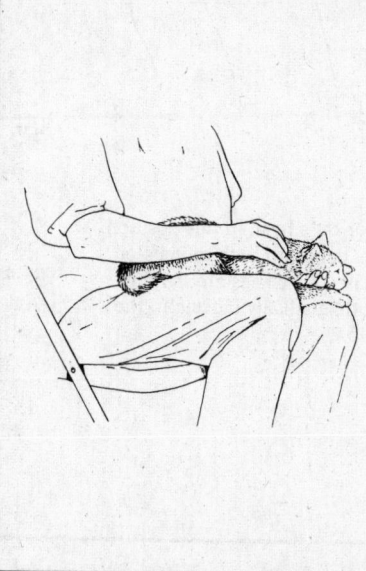

Hold the cat as shown, with the left hand supporting the head under the chin. Alternate the Clouded Leopard and Lying Leopard TTouches, beginning at the base of the neck. Move randomly from place to place on your cat's body and make each circle fairly slowly—in most cases each circle should take one to two seconds. To see what depth of pressure your cat prefers, begin with a number three (see page 26) and experiment with a firmer or lighter TTouch.

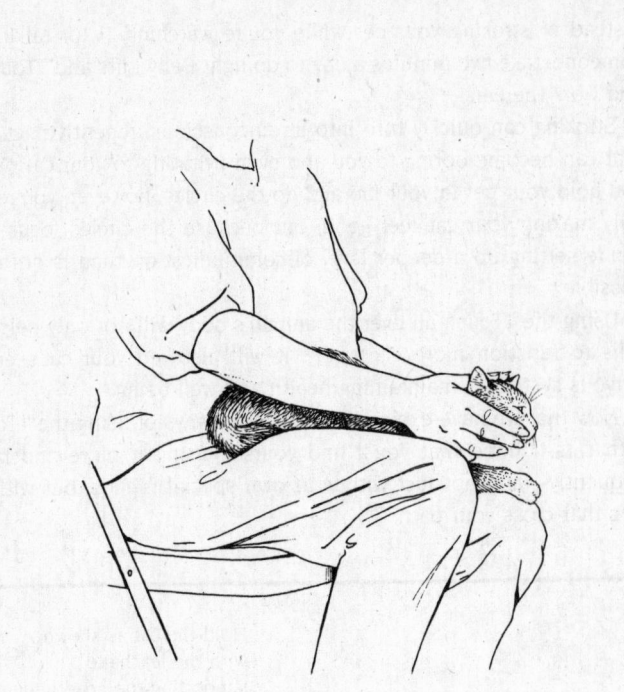

When gently holding your cat's head in this position, you can either stroke her ears with both thumbs or, using your forefinger, make tiny circles around the outside of the cat's mouth, gradually moving to the gums under the upper lip.

The ear has acupuncture points for all parts of the body, so working the ears at random can benefit the entire body. Treating your cat's ears gently, as though they were rose petals, make tiny circles with your forefinger on the inside of the cat's ear. You can also slide your fingers from the base to the tips of the ears, holding them between thumb and forefinger as shown. Not all cats like ear work, so be sure to start out gently.

Support your cat's leg as you use your thumb to make TTouch circles down the leg to the paw. At the paw, gently stroke it, and then make tiny circles around the pads. This is a good procedure for cats who use their claws indiscriminately and as a preparation for nail clipping.

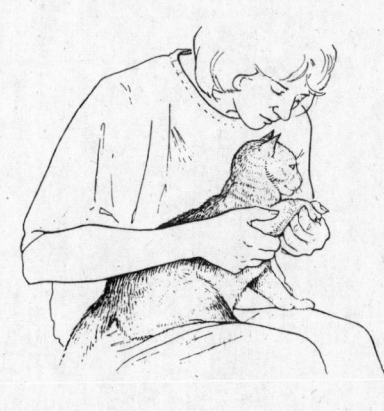

4

Dog Stars

One day I ran into a kind of impasse with Strongheart in which he again became a complete enigma to me. Something was definitely blocking the way. Much to my private embarrassment, I finally found out what it was: ME. With all of my well-intentioned efforts, I had been making the common ego mistake of trying to do all the thinking and arrive at all the final conclusions for both of us . . . I had mentally assigned myself to the upper part of this relationship of ours, because I happened to be "a human," and had mentally assigned him the lower part, because he was "a dog."

—J. Allen Boone
Kinship with All Life

The stargazers of ancient times were history's first movie producers, projecting mythic stories of heroes, gods, and animals onto the vast network of stars glowing in the night sky. Look up on a clear winter evening and there, as part of the longest-running show on earth, you will see the constellation Orion still striding across the sky. Following closely behind him are two smaller constellations: Canis Minor and

63

Canis Major—Orion's faithful hunting dogs. Shining out from the breast of Canis Major is the magically luminous star Sirius, also known as the Dog Star.

Dog and man—a relationship so ancient and so special that it is reflected in the mythology of the sky. How it first began nobody really knows; some biologists believe it started when prehistoric people began taking puppies into their caves for amusement; others believe that over time dogs and men discovered that hunting together was helpful to both species.

We do know that domesticated dogs already existed in neolithic times; archaeologists discovered that an animal similar to the jackal lived with the people of primitive lake villages, and the ancestors of modern sheep dogs have been found in Bronze Age settlements.

The dog family, or Canidae, is composed of thirty-five different species that in addition to our "best friends" include foxes, wolves, and coyotes. All members of this family share certain characteristics like high intelligence and an acute sense of smell and hearing, but the trait that is most meaningful in the history of our relationship with them is their adaptability.

Coyotes, for instance, can travel in a small pack or live the solitary life, can feed on berries or hunt prey. They adapt easily to a variety of climatic and living conditions and are one of the few animal species threatened by humans whose habitats are expanding rather than shrinking.

The first coyote I met personally was named Mindy. Mindy began her sojourn with humans when she was found wandering alone in the streets by a suburban San Diego family who mistook her for an abandoned puppy. Imagine their surprise when at her four-month check-up the vet informed them they had adopted a full-blooded coyote. The family loved her and kept her, but after a while she became too much for them. Not knowing what to do with a coyote, they presented her to the county's animal rehabilitation center, Project Wildlife, for release back into the wild.

But the young coyote was a problem for Project Wildlife, too. Her time spent with humans, with kids and dogs and shopping carts, had created a double bind. She was too "civilized" to be returned to the forest and too wild to make a good pet. She was a refugee—and that is how she entered the lives and hearts of Cheryl and David Nix.

Cheryl and David were well known to the rehab center. They had both been involved with wildlife and animal rescue for years. "No one in our family cared that much about animals, but I was always bringing home snakes, or tadpoles, or strays," Cheryl says. She went on to get a degree in wildlife biology and met David while both were working at a safari wildlife park. David, too, has a degree in wildlife management, and later they established their own wildlife education show, "The Rare and Wild America Show," to present a number of different species to the California public

David and Cheryl were delighted to make Mindy a member of their extended interspecies family of two bears, a wolf, a pair of turkeys, three raccoons, a deer, and several birds, especially since their show had become the official educational presentation of the San Diego Wildlife Park. But Mindy brought a number of special problems with her, which is why, one morning, I received a call from David asking me to come and work with her.

Mindy had come to the Nixes as a one-year-old. Like most wild animals, especially canines, she was territorial and didn't like the shifting locations required by the show. In addition, she had bonded to Cheryl and David, and whenever anyone else tried to enter her enclosure she turned the encounter into a test of dominance. (The expression "top dog" comes from the social patterns of dogs, who have a distinct pack hierarchy.)

Mindy allowed Dave to work with her only when she felt like it. She was still reacting instinctively rather than cooperatively. One of the advantages of the TTouch is that it can be used to teach wild creatures to relate to humans without "humanizing" the animal and without sacrificing respect for individual and species traits. My objective with Mindy was to use the TTouch to take her beyond her instinctual responses to a sense of partnership with humans.

As long as Mindy was on the opposite side of a fence she was very friendly to everyone. She greeted us with leaping enthusiasm, sticking her snout eagerly up to the wire mesh. When Dave took me into her enclosure to introduce us, however, she became nervous and suspicious. So I stood very quietly, giving her space to get used to me.

For a few moments she kept her distance and then approached me and began circling, jumping at my back and lightly snapping and clacking her teeth together, trying to establish dominance as though I was a strange coyote who had entered her territory. I countered by tapping her smartly on the bridge of the nose with my fingertips and she ran away to jump up onto her platform, where she felt safe.

I then approached her with two TTEAM wands (see page 270) and began slowly stroking her with them while toning to her in a low, singsong voice. At first Mindy snarled and bared her teeth, jumping anxiously on and off her platform, but finally she settled down and I was able to move in more closely and stroke her with two lengths of rubber hose (see photo insert), each about a foot long.

When I first approach an animal who is feeling threatened and who is potentially capable of biting, kicking, or striking, I use the wand to maintain distance. But once the animal begins to accept my presence, I use the hose to work more closely and to block any remaining possibility of biting.

By the second session, a few hours later, I was able to work on Mindy directly with my hands. At first she swung her head around to threaten me and bite at the hoses (see photo insert) whenever I approached her shoulders or hindquarters, but then she lifted one back leg in the coyote posture of submission and patiently permitted me to work her whole body from head to tail.

As preparation for teaching her to walk with a lead, I began to restrain her lightly, holding my arms up to prevent her from moving off the platform.

We let her rest another few hours. Then, in the third session, because she hated being confined by a collar, I first took a light rope and draped it over her body, then substituted a towel to accustom her to different kinds of bodily contact.

Mindy loved the towel. She grabbed it in her mouth and we began to play tug-of-war, with Mindy leading me around at first. Gradually and gently, I began initiating an exchange—she would lead me around for a bit and then, when she came my way a few steps, I'd pull her. Then, after a few moments, I'd release the lead back to her again. I never let it get into a wrestling match or rambunctious play, but rather fostered the quiet and cooperative relationship that had begun to develop.

It was like a dance in which switching leads let her know that I understood and respected her concerns about losing control, but was asking her to accept a few things about me in exchange.

I replaced the towel with the rope again, and by the end of the lesson I could lead Mindy with the rope around her neck, while stroking her back and encouraging her with the wand to go forward.

Mindy is very playful and had so much fun with the towel that we decided to leave it with her at the end of the day. With a gleeful look in her eye, she wrestled it around, dunked it in her water tub, dragged it, rolled on it, and finally, in a grand finale, threw it exuberantly up in the air.

As I watched Mindy's exhilaration with her towel, I realized how often I forget to play myself, how "serious" we all become with "maturity"—we imagine we're too busy, too tired, too worried, too old. And we forget, too, that our animals need to play just as much as we do—the cat lying around lethargically in a warm apartment is getting older faster than necessary; the dog eyeing you from his basket is just waiting for a few moments of fun that would inspire you both.

Many of our zoo animals, too, alone and bored to death in their bare confinements, would joyfully welcome the amusement of an appropriate plaything. Cheryl told me that after my visit Mindy's fun and games wore out a number of towels.

On the second day of sessions I brought my friend Richard Champion into the enclosure with me to see if Mindy was still apprehensive about new people. Though she was very excited and we had to quiet her, her previous fearful behavior was gone. She welcomed Richard enthusiastically and even climbed into his lap and stuck her nose playfully into his ear and up his nose, courting him with every coyote enticement at her command.

A year later, the Nixes gave Mindy to the Wild Animal Presentation Program of the Los Angeles Zoo. The zoo was particularly glad to receive her because they had a special need to educate the public about coyotes. Sadly, as cities expand, they encroach on the territories of wild animals, and in the battle for real estate, man usually wins. But in Los Angeles, coyotes, with their adaptable natures, have not moved out of the way. They roam the parkland and wooded hills encircling the city, and because so few people really know anything about them, a great deal of fear, misinformation, prejudice, and controversy surround their continued presence.

Enter Mindy. In her role as a participant in the zoo's educational show she has become a representative for her species, a coyote ambassador to the inhabitants of Los Angeles!

YOUR DOG, THE TTOUCH, AND YOU

▷ *Training and Obedience*

Naturalists call the dominant leader of a pack of dogs the "alpha male," and most dog trainers believe that the only way to school a dog is to play the role of human alpha male. The theory is that a dog perceives the world as a pecking order and will submit to learning only if he is dominated.

Our dogs need to live peacefully in a human-oriented culture. Whether a show dog, a working dog or a beloved companion, we don't want our dogs driving us crazy with unpleasant, annoying, or dangerous behavior. The dog has to learn to live harmoniously with you.

Traditional training techniques, however, often make dog owners uncomfortable or unhappy, because these methods of domination run counter to their special feelings for their animals.

It's hard to take your lively, trusting eight-month-old puppy to obedience class and teach him to lie down by stepping on his collar chain to bring him violently to the ground. It's distressing to learn that the common cure for aggressive behavior is to repeatedly hang a dog by the neck until he passes out, or that a suggested method to end hole digging is to submerge the dog's head in a hole full of water until he nearly drowns.

These methods are all based on working only with the dog's basic instinctive reactions. With TTEAM we approach training from the viewpoint that we can teach a dog to go beyond instinct, that as a complex, intelligent creature he is capable of responding to more than dominance. When we use the TTEAM TTouch we find we are able to speak to a dog's intelligence and his heart; and we can teach him without resorting to force.

Most problems perceived by people as behavioral in nature actually have stress as an underlying cause. Often, what appears to be disobedience or aggression is really a reflection of physical or emotional distress or confusion.

▷ *Aggression: Changing the Hidden Cause*

A friend of mine, Serena Foster, had a Norwegian elkhound named Lars (she also called him "The Wonder Dog") who she thought had suddenly decided to "assume authority over her." As we talked about how we could restore communication between them, it became clear that Lars's problem had as much to do with Serena as with Lars—not an unusual situation between pets and their owners.

Lars was an exceptionally well muscled and beautiful dog, almost as large as a shepherd, with a thick silvery coat, a strong, deep chest, soulful eyes, and a prankster's quick intelligence.

"I don't know what to do with him," Serena said. "He's so loving, but he has this authoritative idea of himself. He challenges me on quite a few things. For instance, he deliberately takes over my reading chair. When I ask him to get down he won't budge. He just sits up really tall, snarls at me, and bares his teeth."

"What do you do then?" I asked.

"Well, at first I yelled at him and he got off, but then the next day we'd just have to go through the whole scene again, so then I tried stroking him and talking to him, but that didn't work either."

"I bet what you really wanted to do was to haul off and whack him, right?" I asked her.

My friend looked embarrassed, and then we both laughed. "It's normal," I said. "Believe me, my first impulse has been just like that in a number of situations, except when the animal was clearly so dangerous that I thought, Uh-oh, better back off, lady.

"It's instinctive to simply react—you want to get in there and growl louder and say, 'Look, I'm bigger than you are,' and whomp him—except he's got sharper teeth than you have."

Serena grinned. "I was afraid of him," she said, "and furious at the same time."

In my experience working with the nondominating methods of TTEAM, this is the exact moment in which you have to stop, catch yourself, and calm down that rush of reactive adrenaline. You have to say, "Okay, now instead of getting angry or frustrated or aggressive or afraid"—and they're all related—"how can I *think* about this? What can I do?"

From what Serena told me I gathered that Lars was intent on controlling her. When a dog sits up tall, holding the tension in his neck and body, he's inflexible. If at such moments you confront him with your own tension, you are only going to stay locked in combat, like two mirrors facing each other.

The TTouch body work is the first step in changing the behavior pattern in such a case. The nonhabitual, nonthreatening motions interrupt the automatic mode the dog is in, and release the tension and inflexibility, allowing him to get out of the loop of reactive response.

To get Lars off the chair for the initial body work, I suggested to Serena that she offer him food. Eating would change his mind-set, defuse the atmosphere of confrontation, and allow him to breathe. Have you ever noticed that you hold your breath when you feel threatened? Holding the breath is a part of "freezing" into a certain behavioral pattern, and releasing the breath appears to reactivate the neural impulses to allow *thinking* instead of *reacting*.

Next, I told Serena to stroke the dog with two of the three-foot TTEAM wands we use specially for dogs. Often, when an animal feels threatened, the direct touch of the hand is too much for initial contact.

I also asked her to keep her body language soft and slow and to avoid looking fixedly at Lars. To create and maintain a mood of calm, I further suggested that she tone to him with words like "Goood dooog, goood boooy."

As pointed out before, toning is a powerful aid. Animal behaviorist Dr. Patricia McConnell, who teaches a course in the human-animal bond at the University of Wisconsin, recorded and analyzed the vocal or acoustic signals of over one hundred and thirty professional animal handlers from around the world (myself included) to see what effect their modulations of breath and voice had on their animals.

"All the handlers did exactly the same thing," Pat says, "whether it was for sheepdogs, sled dogs, horses, or water buffalo. They spoke sixteen different languages, yet all of them used long, continuous, flat or descending noises when they wanted to slow or soothe their animals."

Thus for Serena, toning during her body work on Lars was important—for both of them. The next step was to get Lars to lower his stiffly upright head. When dogs (like many other animals) act reactively, they often tense up and hold their heads high. It seems like a miracle, but as soon as you get them to lower their heads, they give up their resistance, relax, and begin to respond (see page 112). To accomplish this, I suggested Serena stroke her dog's ears first and then work with Lying Leopard circles on his neck and chest.

Once Serena got Lars to lower his head, it would probably be quite easy for her to begin the TTouch on the rest of his body (see illustration, pages 74–79).

Though I was fairly sure that a few sessions of body work with Serena would enable Lars to become a great deal more receptive, the basic question remained: why was he challenging her?

Dogs are emotional creatures just as we are. Like us, they are sensitive to the emotional atmosphere around them and they absorb and reflect the subliminal messages their owners send out. I asked Serena what Lars had been like as a puppy and how he was raised.

"Well," she said ruefully, "Lars was a wonderful puppy, but it wasn't the best of times."

Serena and her husband had bought him for their twelve-year-old daughter when they were in the middle of a disintegrating marriage. They were hoping a flop-eared baby elkhound all her own might help ease her through the family crisis. When Lars was six months old, Serena's husband finally moved out.

"Things were pretty confused and difficult," Serena said. "Lars was my daughter's dog but he was mine, too, and he was so bright and such a strong male. I guess I enjoyed the feeling of being protected that he gave me. I don't think I was all that firm with him." Her smile was a little wistful.

These emotional considerations aside, as a marvelously bright and active young dog just coming into his maturity, Lars was also definitely looking around for challenge. And here is exactly where TTEAMwork with the labyrinth offers a solution (see page 80 and page 268).

When you work with your dog through the labyrinth, you have the opportunity to get him to cooperate without confrontation; you are asking him to listen to you, but your request is in a form that is interesting and challenging to him.

The TTEAMwork in the labyrinth provides an area of compromise, a bridge between you and your dog that allows you to restructure your relationship. Through successful communication, you lead your dog away from the question of who is boss and toward the answer: cooperation.

Serena took Lars through the labyrinth using our basic TTEAM method and the special adaptations that we have found usually work for aggression. The idea is to walk the dog through the labyrinth, constantly varying the number of steps he takes and, alternately asking him to sit or stand while you stroke him with the wand. Pacing with you through the labyrinth, stopping, starting, sitting down, all this focuses the dog and moves him away from reactive, resistant behavior.

Serena began by putting two collars on Lars in case he slipped out of one (I don't like using choke collars). After quietly asking him to sit, she stroked him with the wand, beginning under his chin and going down his chest to his front legs, and then over his feet. This relaxes tension and encourages the dog to lower his head. Next she took several steps with him, asked him to stop, and stroked the wand over the back of his neck, down his back to his tail, then over his hindquarters and down his legs and feet to the ground. This grounds and relaxes the dog and is particularly effective with reactive dogs who very often carry their tension right down into their paws.

I stressed to Serena how important it is to ask for behaviors in a calm and fluid way rather than in clipped, single words that come out sounding like a shot. Try saying "sit" or "stand" or "walk" sharply. Can you feel the tension in your own body? We learned that the most effective way to tell a dog what you want him to do is to frame requests in phrases spoken slowly in a firm yet quiet voice—"Walk on," or "Stand still" or "Sit down." Praising him in the same quiet way, drawing out the word "gooood," also creates a cooperative atmosphere. It's important, too, to be very clear about what you are asking, focusing your own mind so that the dog will mirror your attentiveness.

In addition, your body language will be sending the dog a message. If you stand stiffly upright or jerk the dog on the lead, he will probably echo your tension.

After taking Lars through the labyrinth several times alone, Serena switched to working him together with her daughter, using the Journey of the Homing Pigeon (see page 272). In this exercise, two people, one on either side of the dog, alternate the directions and the stroking as they move through the labyrinth. The Homing Pigeon is especially good for breaking down patterns of habitual resistance because the dog's attention is divided and that keeps him from responding in his customary way.

At first Lars balked a little, hanging back when asked to walk forward, but after a few turns through the maze he got the idea and began to enjoy himself. He looked keen and alert, his ears pricked up, his intelligent eyes eager for the next move.

As a final exercise, Serena tied a body wrap on Lars, bringing it around his chest, crossing it over his shoulders and then under his belly, and fastening it on the top of his back like a Christmas package (see illustration page 264). The containment of the wrap gives them a new and different sense of their own bodies. They feel themselves more, feel the overall wholeness of the body, and this in turn breaks the physical patterns that result in tension, high heads, and growling. With automatic trigger behavior absent, they are open to learning new ways of responding.

We've discovered through many cases that the body wrap is useful not only for aggression but also for working with dogs who are timid, or who are afraid of sudden noises, thunder, strangers, or going to the vet. It's helpful, too, for a show dog who becomes tense when the judge lifts his tail to examine him.

Though Serena began by walking her dog through the labyrinth, with a reactive dog we usually start with the Homing Pigeon and then, when the dog is ready to allow it, we put the body wrap on him.

When I last talked to Serena, she and Lars were doing fine and had reached a new and much happier level of understanding.

"I'd like to try TTEAM on my ex-husband. He's kind of aggressive these days," Serena said, laughing.

"You think you're joking?" I said. "It works on humans, too."

▷ Bud: A Transformation

TTEAMwork can transform the relationship between owner and dog. It can also alter the way in which a dog lives out his life and the way in which he gets along with other animals or pets. By bringing out the best in a dog's character, TTEAM has been known to open some surprising doors, as in the case of Bud.

Bud is a small, graceful collie mix who lives on a ranch in the remote mountains of central Idaho. To get to his home you drive along a steep and bumpy dirt road through sunny upland meadows and deep green forests of cedar and pine.

Many of the nearest villages are ghost towns, remnants of the Gold Rush with names like Granite Creek, Placerville, Grimes Creek, Horseshoe Bend, and Pioneersville. The latter is called Pioneersville now, but during the Gold Rush years, the settlers were so greedy and hoglike about getting rich it was known as "Hogum."

TTEAM practitioner Ann Finley, who lives in the area, told me of a shootout one winter at one of the two gold mines still in operation. Afterward, she said, Pioneersville was much quieter than usual. Locals joked that half the population of the town was in intensive care and the other half was in jail. She flashed me her quick smile.

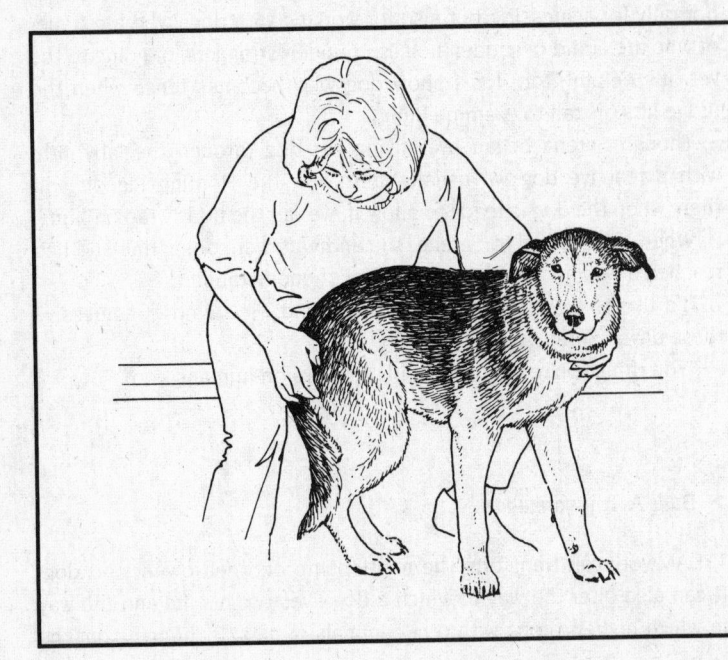

Ann had always been interested in Bud. He lived at the ranch neighboring her own thirty-acre spread, and whenever she went to visit there she felt sorry for him. Bud was miserable. Constantly in trouble, he spent most of his days on the end of a chain.

Dismissed as too "hyper" to be of any use with the cattle, he slunk around, nose to the ground, running with such a rapid, fluid motion that people used to say he was part coyote. He could not be touched or petted, except by his owner, Xinia Jones, and whenever he was let off the chain he chased her prize-winning rabbits or was himself bullied by Rover, a stocky brown-and-white mutt who was the top dog at the ranch.

Ann had read about the Tellington TTouch in a magazine and she began applying what she had read to the job of befriending and training a colt she was rearing. Amazed at the depth of communication opening up between her and the animal, she started attending my clinics and soon became a familiar face.

The TTouch on the hindquarters is particularly beneficial for dogs who are fear-biters, overly timid, or afraid of thunder and lightning or gunshots. Some dogs will be concerned when you work this area and will try to sit down. Should this occur, lighten your TTouch and reassure your dog with your voice. Asking your dog to stand while working the hindquarters helps to release the fear more quickly than if you work while he is lying down.

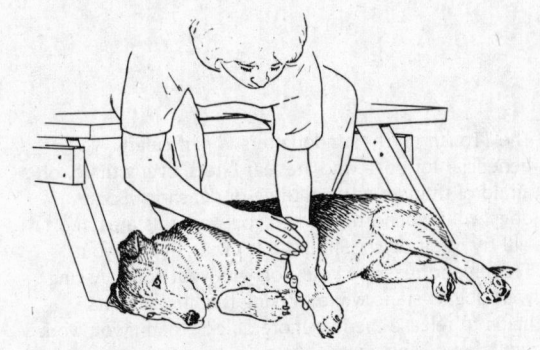

If necessary, you can also work on your dog while he
is lying down. For some nervous dogs, this can be
threatening at first, so gently but firmly keep your dog
quiet with your right hand as you make the circles
with your left. If a dog is overactive or very submissive,
he may lick or try to get up, and it may take a few
lessons for him to learn to focus and relax.

This dog is relaxed as the handler uses the Clouded
Leopard on his front leg. Notice how she supports the
leg with her left hand so she can use the TTouch on
his shoulder, leg, and all the way down to his paw.
Using both hands, you can gently move the shoulder
forward, up, back, and down, and then reverse. This
movement is not meant as a stretch but rather to
encourage relaxation and easy range of motion.

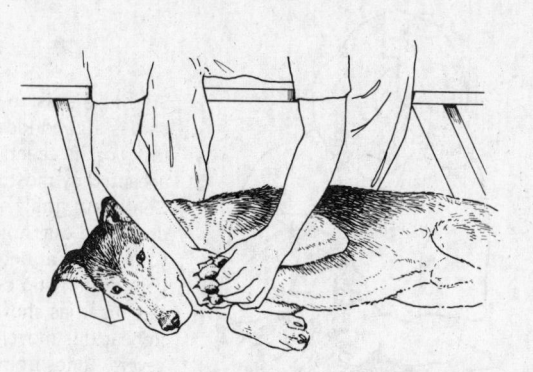

Here the handler works the paw pads using the Raccoon TTouch. This area can be very sensitive, so approach carefully. It's best to start the TTouch on a less sensitive area and then work your way to the feet so that the dog has been prepared to relax his paws. This work is especially helpful for older dogs with heart problems.

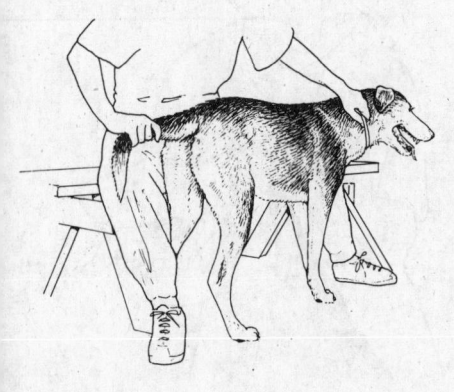

Lift your dog's relaxed tail and, following a straight line from his body, slide your hand toward the end of the tail in a series of pull-and-hold movements, each lasting four to six seconds. Hold your dog by the collar to help keep his body straight.

Starting a session with the ears is a good idea, since this work is eagerly accepted by most dogs. Gently but firmly slide down and out, holding your dog's ear between your thumb and bent forefinger, as shown. Repeat this movement several times from the base of the ears to the tips, covering a different portion of the ear with each slide.

Make tiny circles as you slide from one place to the other all over the ear. The thumb provides a base while the forefinger circles, or vice versa. Robyn used this TTouch on her old dog Stash, who was fourteen. He had fluid around the heart, and would often have trouble breathing and sleeping. After Robyn worked his ears for a few moments, he was able to get comfortable and drift off to sleep.

Sit so you are behind the dog's head and supporting his muzzle with one hand. Be careful not to squeeze, which could cause him to fight. Use the index finger of your other hand to carefully lift the upper lip. Gently rub the inside of the lips and make tiny circles on the gums. If the inside of your dog's mouth is dry, wet your fingers with warm water. If your dog is unhappy with this work, begin by making very light circles on the outside of his dewlap before you slip onto the gums. Quiet perseverance over several short sessions usually brings acceptance.

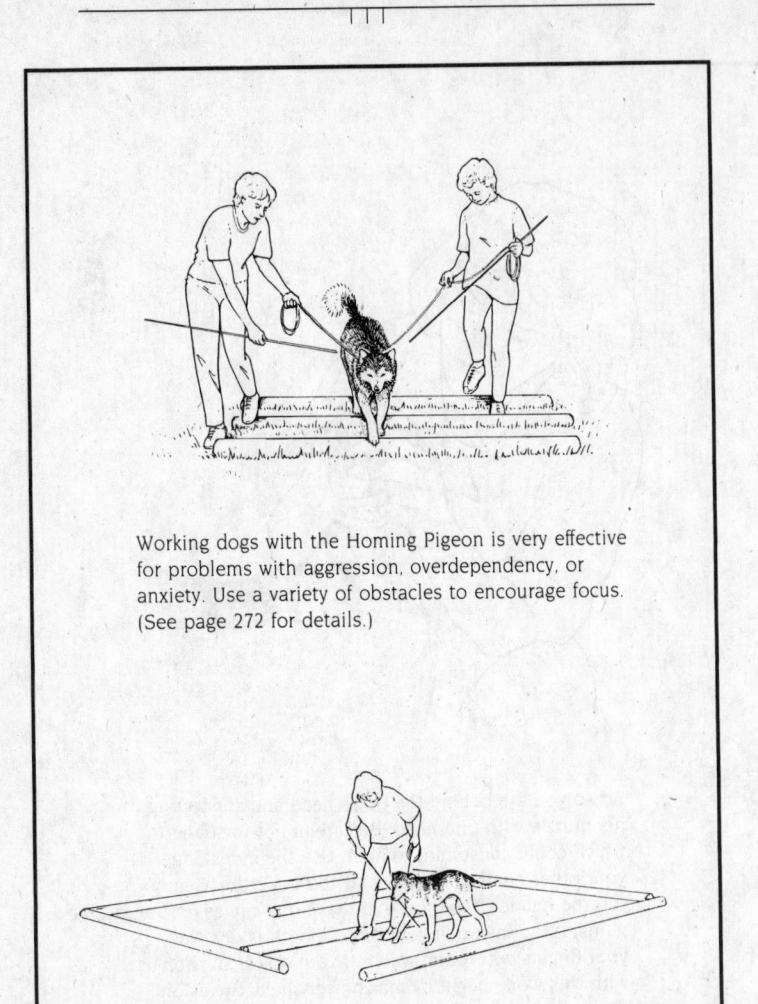

Working dogs with the Homing Pigeon is very effective for problems with aggression, overdependency, or anxiety. Use a variety of obstacles to encourage focus. (See page 272 for details.)

Taking your dog through the labyrinth: Use the wand to slow the dog down by touching him on the chest and moving the wand out in front of him as a visual aid. The wand helps the dog to focus, rebalance, and listen to you.

Bud, of course, was a perfect candidate for the Tellington TTouch. But not only for TT. Ann had been waiting to find a special dog for a teaching program she was working on, and she had a hunch that her search had ended with Bud. Interested in exploring the world of education, she had graduated from Arizona's Prescott University with a degree in human development and had also worked for nine months teaching at the Pine Ridge Indian Reservation in South Dakota, where, she says, she learned as much as she taught.

When she came to study with me I had just established Animal Ambassadors International, a multifaceted, global organization whose primary goal is to increase awareness of the importance of animals in our lives (see page 223). Among the many projects that AAI fosters for this purpose is the Animal Ambassadors School Program, first introduced to the New York elementary school system by Alexandra Kurland, a TTEAM practitioner and well-known children's-book author. One of the main aims of the program is to show children that communicating with animals and becoming their protector can make a real difference both to a child's own life and to the world.

Each student chooses a specific animal—bear, eagle, crocodile, lion, snake—as his or her own special totem and teacher to learn about, cherish, and defend. The children also learn how to communicate better with their own pets through practicing the TTouch on special animals who come to visit them as ambassadors.

Ann fell instantly in love with this project and decided to bring live animals to the children. She set up a pilot program in a school in Boise, and when it came time to elect an animal ambassador, she couldn't stop thinking of Bud.

Xinia was delighted at the idea of lending him to the experiment. Not only was there a chance that he would stop eating her rabbits, but perhaps he would even become a star.

The only problem was that Bud was not sure he wanted to be in the spotlight. When Ann came to see him for her first working visits, he had to be coaxed from his hiding place under an old couch that had been put out in a corner of the yard. Once the dog had succumbed to the lure of a dog biscuit, however, Ann began work by going over his entire body with the Raccoon TTouch.

At first the collie found the contact worrisome and wondered nervously what was expected of him. He trembled with apprehension and could tolerate only the lightest of pressures for short intervals. After several sessions, as the TTouch calmed and focused him, he became less scattered and threatened. Finally, after four sessions over a period of several weeks, to both Ann and Bud's great excitement, he broke through his fear and began to respond, enjoying the communication during the sessions and greeting her happily when she came to see him.

It was time to take him to school.

"Although I was sure he would do well, I wasn't prepared for the amazing way his character blossomed," Ann told me. "He was still shy and sensitive but he was a born communicator."

Bud liked to read character. Each child who worked on him received an individual greeting. Sometimes it was a single swipe of his tongue across a cheek, or a gentle nuzzle on the hand with his nose. It could be a paw shake, or even just a penetrating gaze. Once, when a child grabbed his proffered paw and tried to pump it up and down, he retreated for several minutes, teaching the children that he wasn't just playing and that if you want to talk to an animal you also have to listen.

Of course I was delighted to hear about Bud's success in the Boise school system, but I was also very curious about what was going on "back at the ranch." "What about his rabbit habit, what about Rover?" I asked Ann.

She laughed. After she began doing the TTouch with him, she said, he no longer slunk around furtively but sparkled with the proud confidence of his true collie nature. His newfound assurance, however, also triggered unexpected strife.

Bud's transformation had freed him from doing time on the chain, however his pleasant life was not secure as long as Rover remained alpha dog and king of the castle. Feeling strong and assertive now, he turned on Rover and challenged him regularly and fiercely.

"Poor Rover was completely confused by this 'new' Bud and his constant attacks," Ann told me. "But Rover was much heavier and more thickly muscled than Bud and beat the socks off him every time."

Ordinarily, a dog who loses a challenge to the alpha dog of a pack will acknowledge defeat and withdraw. But not Bud. Instead, he simply moved on to the last phase of his already remarkable transformation.

Instead of clashing with Rover, Bud was finding ways to avoid confrontation without losing face. If he wanted to go somewhere and Rover was in the way, he would work out a method to do it without provoking the other dog. Xinia pointed out to Ann with amusement how, instead of crossing Rover's path, Bud would always manage to put something or someone between himself and the other dog, cleverly creating a buffer zone. Ann would laugh, watching him as he made his maneuver, his body language elaborately casual, as if he had no idea that anything out of the ordinary was going on at all. But she noticed that his trot would be especially jaunty and his plume of a tail held high.

And so it should be—Bud is a winner. Not only that: Xinia says that Bud is now the best cattle dog on the place.

▷ *Aggression, Pain, and Irritability*

I first met my literary agent Reid Boates on a wintry afternoon in New York City. We sat around a table with cups of tea and examined photographs of me working with snakes, elephants, bears, and coyotes, with ferrets, parrots, snow leopards, and dolphins.

Reid put his cup down to look at me. "Amazing," he said.

"It's not *me*," I told Reid, "it's the TTouch. The fact is, anyone can learn it, although of course I don't recommend starting on tigers."

He raised his dark eyebrows quizzically.

With skeptical people, I always use the hands-on approach. They may not believe what you say, but they can't deny what they experience. Almost everyone has a little ache or pain that bothers them, and with Reid it was his knee. "Watch how simple it is and how good it feels," I said, and coming around beside him I demonstrated what five minutes of the Raccoon TTouch could do for him. Within minutes he was asking me to show him how to do the circles on his somewhat hyper Airedale, Maggie.

Two days later, he called me excitedly with a report, but not about Maggie.

The night before, at a dinner party, he had met a very unpleasant little dog. The party was held at a New York town house opposite the Metropolitan Museum by a client of his, the author of several best-selling cookbooks. This was to be a special tasting of a selection of brand-new recipes for cold weather.

He entered the high-ceilinged drawing room to find twenty or so guests chatting nervously over their drinks and hors d'oeuvres, pretending to ignore a scrawny, ancient little terrier yapping away at floor level.

Reid opted to use the "facilities," which were upstairs, before he joined the battlefield. But as soon as he set foot on the bottom step, the feisty little creature spotted him and hurtled across the room to nip nastily at his heels. The next instant the dog had run up the stairs in front of him and sat blocking the way with a surly look.

"Now or never," Reid thought. Reaching out his hand, he touched the dog on his shoulders and rather thin breastbone, starting out with the tiny light circles of the Raccoon TTouch he had learned from me a few days earlier.

At first the dog looked at Reid with suspicion, as if to say, "Don't fool around because you've got about one minute to make this work." But then, almost immediately, he began to soften. His eyes, which up until then had been focused on Reid's heels or other nippable places, lifted and cleared as though he had suddenly woken up and found himself in a different world. He sat motionless, raptly attentive, keeping eye contact with Reid, who continued to make circles across a wider and wider radius of his chest.

"No, no. What are you doing? Don't touch him there," a shocked voice called out, and Reid turned to see the horrified face of his hostess. "He hates that," she said. But by now the dog was utterly entranced; she just stood and watched in amazement.

Finally, when Reid stood up, the dog followed him thankfully into the drawing room and lay down at his feet.

The poor terrier, Reid learned after discussing him with his astonished hostess, was suffering from severe arthritis, mostly in the breastbone.

There is no question about it: prolonged, unrelieved pain makes animals as well as people irritable and aggressive. If your dog is unusually belligerent and snappish, go over his whole body with the TTouch and look for painful areas. At the beginning, you might find it useful to note the areas where the dog is nervous about certain types of circles and where he is not. You'll discover that over a period of several days the areas that you can touch without nervous response will enlarge. For instance, a dog that will not let you touch him on the hindquarters will slowly begin to accept the TTouch, and will cease to sit down or growl at you each time you try.

Sometimes a TTouch examination will reveal unexpected sources of a dog's pain and abnormal behavior. For example, I was once consulted on a case in which a rottweiler had several times threatened both his owner and his trainer. Rottweilers grow to be around one hundred pounds, so even though this fellow was only six or seven months old, he was already pretty formidable, with his big, broad head and sturdy body. The rottweiler breed, whose ancestors were once used to drive cattle over the Alps for the Romans, is especially renowned for bravery and calmness, but this pup's genes, it would seem, hadn't received the news.

After several incidents, the trainer resorted to the traditional way of dealing with aggression: he lifted the dog up off the ground by the choke chain and kept him hanging there until he passed out. The owner was upset, and when a mutual friend and TTEAM practitioner called to discuss alternatives, I suggested that work on the dog's mouth might bring about a change in his emotional response (see pages 116 and 266).

"Yes?" my friend said, with understandable nervousness, "and just how do I get in there?"

I reassured her that if she followed the steps of the TTouch slowly and carefully she'd be doing just fine. Start with the soothing strokes of the wands first, I told her, and then using Lying Leopard circles, work the whole body including the head. As it turned out, she was able to skip the wand work. The pup was admirably amiable as long as he wasn't being asked to do something he didn't like.

However, as soon as she reached his head, a deep, threatening growl would start rumbling up from his throat. Not to worry, I said, the next step is to use both hands at once. While the left hand works the places on the body where the dog had actually enjoyed the TTouch, start working the right hand up onto the tense area of the neck. With the right hand, circle up the neck and onto the head as far as you can until the dog lets you know he's uncomfortable. Then stop, go back to the point you reached just before he growled, and begin again from there.

Using this technique of the left hand giving reassurance while the right hand inched farther and farther forward into the red-alert zone, my friend was able to reach the mouth in two sessions, unscathed by even a single nip. (See mouth work illustration page 79.) And then came the kicker—once she actually got to work on the dog's mouth she made an astonishing discovery. Still in place behind the dog's full set of adult teeth was a complete set of baby ones that had yet to fall out. The poor animal was walking around with two sets of teeth crowding his mouth, which most certainly must have been uncomfortable and painful. No wonder he was irritable and easily riled.

My astonished friend reported her discovery and then thoroughly worked the mouth area, making small circles over both lower and upper gums and gently manipulating the dewlaps. A month later the dog began losing those troublesome baby teeth and there were no more reported difficulties.

Chewing the House Down

It was 1980 and I was in Seattle to give a six-week course on TTEAM to a group of European horse trainers and riders. The fifteen visitors were from Belgium, Germany, and the Netherlands, and we all lived together in two big lakeside houses on the outskirts of the city.

In the evenings we'd gather on the wooden decks that stretched out over the water and talk about what we had learned during the day. Then we'd head happily off to sleep. One night, just after everyone had drifted off to his or her room, I received a call from my old friend Nancy Navarra. She lived nearby and suggested I stop in and say hello. Nancy is a Feldenkrais practitioner and we always had an exciting time exchanging ideas.

I drove to her small cottage on a quiet suburban street and found a new addition in her life—Blitzen, a big, bouncy, six-month-old German shepherd, all wagging tail and eager eyes. I immediately noticed that the legs on all the chairs and tables had been gnawed without mercy and that Blitzen had been busy with a wicker magazine basket, too. The house looked as though a beaver had come through it.

"I love him, but he's chewing the house down," Nancy said mournfully. "It seems to be the only way he can relate to anything, including people. I can't even leave a sweater out. He chews up the armholes like a giant moth." She sighed.

It would have made me laugh if I hadn't known how truly frustrating it was. Nancy had tried everything: spraying the furniture, smacking the dog, making him sit, hitting him with a rolled-up newspaper, ignoring him, trying to push him away, and even taking him to the vet. But nothing stopped him.

He's teething, I thought. At that time I had not yet had the opportunity to try the TTouch on very many dogs, so I was happy to have the opportunity to work with him. I moved along his body following the flow of my intuition, allowing my hands to lead me where they would. Finding myself repeatedly drawn to Blitzen's mouth, I began to work his gums with tiny, firm Raccoon circles and on his lips, tongue, and palate with light Lying Leopard circles.

Following your intuition in using the TTouch is a primary part of the method. Though you may know the exact pressures and hand and finger positions for every TTouch, only direct contact with your animal can tell you how, where, and when to use them. You just have to trust and let go of controlled thought—I call it "letting your fingers do the walking."

In those early days of the TTouch I didn't know that you don't necessarily see a change in the animal's behavior at the time of the session, nor had I discovered that a session of ten minutes was usually long enough.

So I overdid it, giving the poor dog's mouth a forty-five-minute workout while I became increasingly discouraged at the lack of results. Not only did he continue to try and mouth me, but he hated what I was doing. Blitzen was a sweet dog and didn't try to bite, but he struggled and constantly tried to pull away from me. And I, of course, determinedly continued on. Finally I went home, convinced that my efforts had been useless.

I was so sure that the session had come to nothing that I didn't even ask Nancy about Blitzen when I talked to her again. To my amazement she said, "Guess what? Almost immediately after you left the pup stopped chewing anything off limits, and he hasn't put a single toothmark on anything since."

Now, after years of working on a variety of species with problems like chewing, biting or aggressive behavior, I find that you can often trace these difficulties to something that is physically disturbing in the mouth, like teeth that are just coming in, or two sets of teeth, or a mouth that is really irritated.

I remember working with a horse, a two-year-old who was very explosive and unmanageable. Everyone thought he was just plain crazy until I found that, like the rottweiler, he had two sets of teeth. I also remember another three-year-old filly who had always been a lovely horse until suddenly, out of the blue, she began attacking her owner. I examined her teeth and found that her gums were distended, hard, and had a strange white color. We found out subsequently that she was suffering from cancer.

▷ *Barkers*

The usual way most of us deal with the irritation of constant, unwarranted barking is to scream "shut up!"—or words to that effect—at the top of our lungs, or, alternatively, we simply "tune out" the sound. While the shouting "'technique" may gain you silence for a while, it's strictly a limited reprieve. After several hours, Rover will more than likely become alarmed again and the whole process will begin anew. By this time you are thinking evil thoughts about Rover and the neighbors are dreaming of hiring a hit squad for both of you.

It doesn't have to be that way. Contrary to what you may be feeling, your dog isn't just being neurotic or annoyingly hyperprotective. Nine times out of ten when a dog barks at every little thing and even seemingly at nothing, he is compensating for a feeling of insecurity. Because of his lack of confidence he's afraid and full of the muscular tension that fear causes.

A dog who is nervous and barking will hold his head stiffly erect and tense his hindquarters, blocking the neural impulses that allow him to sense the rest of his body and thereby undermining his self-confidence. So, in a vicious cycle, he barks because he feels insecure and he feels insecure because he's barking. To help him, we can focus our TTouches in those places we notice tension.

Dogs who bark excessively often carry a lot of tension in their mouths. In a quiet moment, begin to work with the mouth by cradling his chin in one hand from behind while you gently stroke his lips and muzzle with the other. Then begin to work inside his mouth by touching his gums in a circular motion. You may need to wet your finger. A good deal of emotional stress is carried in the mouth, so you may only be able to work for a few seconds before the dog pulls away. Short sessions work best.

Dogs who bark incessantly may be frustrated that we are not listening to their pleas. They may be desperately trying to tell us something is going on outside, and repeated dismissals of their efforts may erode trust in us. Many dogs simply want their alerts to be acknowledged, "Thank you, Fluffy—it's just the recycling truck."

Many people are puzzled when I suggest reassuring a dog when he barks. "Why are you telling the dog he's okay when he just did what you don't want him to do? Doesn't that give him the message that if he barks he'll get attention?" they ask.

The answer is—don't make him have to bark to get attention. Give him quality attention when you're with him. When you're sitting peacefully with him, use the circles rather than just idly petting him. Tell him how good his quiet attitude is. As with cats (page 59) it will focus and intensify your communication. When your dog starts barking, reward him the minute he stops and give him plenty of praise and reassurance. He will soon learn to understand that *not barking* is the desired behavior.

TAKING THE TRAUMA OUT OF
BATH AND BRUSH TIME

For several lucky years I lived and traveled with a really unusual little dog named Bonnie. Often a cat or dog will simply walk into your life and claim you, and that's exactly what Bonnie did with me.

I was visiting the stable of a friend, Jessie Bradley, looking for a horse to buy for a student, when I saw this little, black, moplike critter hiding in a corner. You honestly couldn't tell the nose from the tail unless you lifted up her long scraggly hair.

"What is this?" I asked Jessie.

She sighed. "The poor thing jumped into our car at the shopping mall a week ago. We placed ads and put up notices everywhere, but no one claimed her, and since we can't keep her anymore we're going to have to take her to the pound."

"Oh, I'll just take her for a few days until I find her a home," I said. Of course, everyone but me was aware that she had already found a home—mine.

I decided that she needed the professional ministrations of a good grooming establishment. "Oh, my God," I gasped when I saw her two hours later, shocking her so much that she ran off and hid in a corner. The groomer's clippers had revealed that my Cinderella mop-dog was a princess in disguise. Her true identity? A Scottie of apparently impeccable lineage.

Bonnie never minded grooming, but many dogs have to be dragged out of the depths of a convenient closet once they sense that you have a session of health, hygiene, and beauty on your mind.

Let's begin with brushing. Every animal and every human has a different level of sensitivity to it. For example, try taking small sections of a friend's hair at a time and making small circles with it. Exchange places to see how you respond yourself. Then try the same thing with another friend. You'll be amazed at the differences; some people will want their hair pulled quite firmly, while others will find anything but a delicate butterfly touch painful.

If a dog is sensitive, she may have become afraid of the brush. Be responsive rather than dominating; for instance, if she runs away every time she sees the brush or is reluctant to get on the table, cringes, or lies down, put the brush down and then bring the dog to the table. Or if the dog is still very nervous, take her to a place where she is used to having a good time with you. Give her a little food treat and begin doing the TTouch.

After a while, run the back of your hand (less "electric" than the palm) over the dog, instead of the brush. Or take a very soft material like sheepskin and use it to make slow, light strokes and circles over the body. Tone as you do this and really watch your breathing.

One of the reasons for toning is that it slows your own breathing and movements, so when you take up your brush, tone and breathe to help you keep your strokes slow and rhythmic. When you're brushing a dog who is sensitive, I think it's really important to breathe in rhythm with the brushing. So often when people have trouble with grooming it's because they start off banging away at the dog. He's nervous, and they're trying to get it over with as fast as possible. Of course that only makes the dog more nervous. It's like the difference between having someone brush your hair with hard erratic strokes or long rhythmic ones in soothing harmony with your own breathing.

So if you soften your shoulders and body when you brush, and stroke with rhythmic repetitions, your dog will be able to breathe naturally and won't tense his body against the brush. Try to find the places where the brushing feels good to your dog. If she brings her head up, ask her to lower it (relax tension), or lie down. With a nervous dog, divide the necessary time up into three- or four-minute segments rather than an extended twenty-minute session.

Similarly, bath time doesn't have to be a wild struggle. Instead of just grabbing your dog and throwing him into the tub, take the time to create a cooperative atmosphere with a preliminary session of basic TTouch. The session doesn't need to be longer than a few minutes. Begin with Python Lifts (page 254) and ear work for relaxation, then go on to the Clouded Leopard (page 241) to bring your dog to a state of calm focus. Then, whether you are bathing him in the tub, on the floor of the garage, or outside on your lawn, make sure that the water temperature is moderate and that you wet him gradually rather than all at once.

Don't wet the face first; begin by wetting the legs from the paws up with a bucket or large kitchen pot. Talk to your dog, acknowledge how good he's being, and don't try to hurry through it. If you take the time to empathize with your dog, you'll find that at the end of the bath you'll have more than a clean dog—you'll have a better relationship with him, too.

VISITING THE VET

Many dogs hate going to the vet—they hyperventilate, won't sit still, and some become aggressive and hard to manage. These symptoms of fear may have varying causes. Perhaps the dog remembers an unpleasant experience, or perhaps he is picking up a sense of illness and anxiety from other animals in the office.

Recently, a woman brought her dalmatian to me because the dog was afraid of thunder. That wasn't all, as I soon found out. We were working the dog in the Journey of the Homing Pigeon. When I stroked her on the hindquarters with the wand the dog tucked her tail and growled.

"That's exactly what she does at the vet," the woman said exasperatedly.

"Good," I said. "Then here's the chance to get her over it." So we simply continued taking the dog through the labyrinth, starting by stroking the wand on the neck and shoulders and then going briefly to the hindquarters and tail and repeating the movement. As we went through the labyrinth, we stopped to reward the dog with nice, big Abalone circles on the shoulders, and varied our movements as described on page 80.

After several sessions of fifteen minutes or so, a gleam of pleasure began to awaken in the dog's eye. She was beginning to enjoy herself.

A peaceful trip to the vet is the result of an educating process and attitude that has to be in place well before you get into the car for the trip. So what does "educating" mean?

It means teaching a dog to accept new experiences with interest rather than fear. It means reaching a level of communication with your dog that allows him to trust your intentions for him completely. It means educating yourself, too, learning to approach challenge quietly so that your dog can mirror your confidence instead of your apprehension.

Once you've worked with your dog using the TTouch body work and the labyrinth, he will be accustomed to responding to you with trustful cooperation rather than reflexive submission. In situations that make him anxious you will be able to say "Now let's just slow down and think this over." Obedience training and show dog exercises that emphasize dominance and demand get desired results as well, but in many cases, while your dog may be well behaved, he will also remain tense.

To build up your communication use the TTouch frequently. Make circles on your dog whenever you can, whether you're just sitting around, watching TV, or chatting on the phone.

Then, with this foundation of connection and training as a base, nine times out of ten your dog will need nothing more than a little refresher course in the labyrinth (in your yard or in the park) to be the perfect traveling companion, to the vet or anywhere else for that matter. (The same techniques are successful in combating car sickness. Prepare your dog with sessions before you travel and a few minutes in the car before you set off.)

HYPERACTIVE BEHAVIOR

A friend of mine is a zoo director. He knows animals really well and has a calming effect on them, but like the shoemaker's children who have holes in their shoes, his own dog was uncontrollably hyperactive. For example, each time he came home, Sparky, a wire-haired terrier, would greet him as though he had just returned from a twenty-year absence. She practically knocked him off his feet, jumped all over him, and raced around the house barking incessantly.

Dogs like Sparky, who have a somewhat hysterical temperament, need a framework of consistency and firmness—otherwise living with them becomes an ordeal. When my friend asked my advice about the terrier, I suggested he start by teaching her to sit quietly when she greeted him. When a hyper dog is permitted to go crazy about one thing, she will do so about every new experience.

It's possible to greet a friend joyously without having to go into hysterical paroxysms, but somehow we equate an overexcited response in our dogs with love and happiness. To train a hyper dog to calmness we first have to retrain ourselves to understand that we get just as much emotional response from them when they are quiet and connected to us through the TTouch and through eye contact as we do when they are leaping around like fish out of water.

Hyperactive behavior can also express itself as a constant need for attention. On a television show in Germany I demonstrated how to make the single circles for nervous animals (page 44) all over the body of a little dachshund who was a fear biter and snapped when stressed. The dog was very high-strung and defensive and I showed how the circles would give him more confidence. (Most people think biters need less confidence but actually the opposite is true.)

A week later I received a phone call from a woman who had seen the show and had decided to try the method on her Irish setter. The dog was very nervous, and every day when the woman came home she would sit down and lavish attention on him. The trouble was, it was never enough. Each time she stopped petting him, the dog would nudge the woman's hand with his nose for more.

Thinking she had nothing to lose in trying the circles, she did three minutes of the TTouch on her setter, after which, to her surprise, she was able to get up and go about cooking her dinner without feeling guilty, while the dog lay quietly watching, apparently satisfied.

This seeming miracle works for two reasons: One is that the circles activate the dog's awareness at a cellular level so that unlike stroking, which leaves no imprint, the experience remains with the dog even after you've stopped petting him.

Secondly, by making the circles you have changed the quality and type of attention you are giving your dog. Slowing yourself down and breathing into the circles means that you yourself are relaxing. Instead of patting the dog while your mind is on making dinner or whatever it is you want to do next, you become focused and grounded and so make a *real* connection to your dog. Animals are wonderfully honest—the only way you can truly connect to them is by being fully present.

YOUNG PUPPIES AND OLD-TIMERS

Martha Jordan is a TTEAM practitioner and passionate conservationist who wears many hats with equal skill and ease. Licensed as an animal rehabilitator and massage therapist, she is also a wildlife biologist, has had wide experience as a dog trainer, and is an expert on wild geese and swans. One of her favorite jobs, herding and banding wild geese for the United States Department of Fish and Wildlife, combines several of these skills and interests.

For her first trip into the Yukon National Wildlife Refuge in Alaska, Martha took along her Border collie, Tucker, whom she had trained to be her partner in herding the geese. "Tucker was a wild and woolly dog at first," she reports, "but we found that TTeam ground exercises really taught him to focus and listen.

"He loved his work," Martha says, "but it wasn't easy. The area was flat tundra and very watery, mostly rivers and lakes—and very muddy," she adds, laughing.

"We would fly in to a river with two float planes, set up nets blocking the way, and then take off again to find a flock of geese. Once they were located, we chased them toward the nets in a three-way herding partnership, one plane in the air, one on the river, and me and Tucker on the ground."

Since her Alaskan roundups Martha has assisted in training a number of dogs for this work, and the program has expanded to include the Aleutian Island chain, where the roundups are used not only for banding the geese but for relocating them when necessary. The dogs are invaluable, Martha tells me: with the addition of only one dog, a roundup of close to two hundred geese that once took four weeks can now be accomplished in four days. And the idea seemed to be catching on. Word about Martha's goose-herding techniques spread to the former Soviet Union, where the government expressed interest in her work.

As a former breeder and dog groomer, Martha is still involved with a variety of dogs of all breeds and ages. The puppies of a number of breeds have their tails docked at an early age and I suggested that Martha work the TTouch on their tails for a day before and two days after the docking. Docking disturbs a pup's gait and equilibrium, and the TTouch makes a big difference in his adjustment to his new condition, helping him to quickly regain balance and confidence.

We have found that puppies that are born prematurely or damaged can often be helped by five-minute sessions of very slow, very light Clouded Leopard TTouch. You can see them becoming more alert and vigorous right before your eyes. The TTouch seems to give them a jump start on life. Frequently, they begin to move more confidently and with more interest in their surroundings than is usual under the circumstances.

Working puppies' ears and making tiny circles on their lips, gums, and tongues can help to activate the sucking reflex and start them nursing, or aid them in nursing more vigorously. Small circles around and under their tails with a warm, damp sponge imitates the way a mother dog usually licks her babies and stimulates them. This is especially necessary if you have a first-time mother who doesn't know what to do for her pups herself.

The TTouch is also helpful with problems of aging. Bonnie Reynolds first discovered the TTouch when her Australian cattle dog, Mr. Frazer, was sixteen years old and near death. A best-selling author, she has since given up writing in order to work with animals. Bonnie is a TTEAM practitioner and also runs a nonprofit retirement home for aging horses in upstate New York.

Mr. Frazer's full name was actually the Right Honorable Sir Malcolm Frazer, after the Australian prime minister, and Bonnie thinks he did the name proud. She found him while she was visiting Canberra in Australia. "I had just ended a relationship," she says, "and was walking unhappily out of my hotel, when I saw this desperate creature curled up on the cement outside the door. Obviously someone had kicked him out of a car."

So of course Mr. Frazer jumped right into Bonnie's car, and into her heart as well. They were inseperable from that day on until he died at the age of nineteen.

Australian cattle dogs are a mixture of wild dingo, Border collie, and blue heeler, and it's from the heeler that they inherit both their slate blue coats and their fantastic ability to herd (they round up sheep and cattle by nipping at their heels to keep them moving along, hence the name). It's like watching a great athlete in action to see a heeler avoid the kick of a steer; suddenly, in the midst of a full run, he will drop and flatten himself to the ground, just milliseconds before a hoof goes flying harmlessly over his head.

Bonnie brought Mr. Frazer home to California with her where he caused quite a stir on the beach at Malibu. His dingo blood gave him the look of a coyote, and people were always stopping her and asking what she was doing with a blue coyote. When I met Mr. Frazer, he was having the kind of problems you might expect in a dog sixteen years old.

"He would wake up in the mornings and just lose it," Bonnie remembers. "He'd start staggering around and falling, and finally he would just fall over and lie there looking at you, and he was not going anywhere. He couldn't get up."

Bonnie didn't want to give up on Frazer. Using the TTouch she began working on his legs. She went over every inch of all four legs from shoulder and haunch to foot, using as she says, "hundreds and hundreds of tiny Raccoon TTouches," all the way down to his feet, where she continued to make the tiny circles on his paw pads and on the webbing between the pads.

To her amazed delight, after forty-five minutes of this concentrated TTouch treatment, Frazer would get up and frisk around like a puppy. "He was happy because he didn't give a damn about anything but being with me," Bonnie says. "He was the love of my life, and the TTouch kept him going for an extra three years."

As it did for Mr. Frazer, daily sessions of TTouch body work plus the tiny Raccoon circles on the legs and feet should help immensely in prolonging the life, zest, and overall stamina of your aging dog. Not only is ear work particularly beneficial for older dogs, but they generally love it, and if your dog has problems with his lungs, we've discovered that Abalone circles (see page 248) on the chest area can relieve labored breathing.

Don't think that because your dog is old and sleeps much of the time he has necessarily lost his love for play and his desire to be with you. Even if he can't dance around the way he used to, he will probably light up if you bring him a favorite old toy and play with him. If he can't run well anymore, make adjustments and conduct your play on a smaller scale. Or if your dog no longer has the desire or energy for play, you can keep up your closeness and the quality of your connection by doing the TTouch body work on him.

It's important to maintain your contact with him even though the range of your dog's mental and physical abilities may be diminishing, because surely his love for you continues to shine somewhere inside him, as bright as the Dog Star, Sirius, who never stops following the footsteps of Orion across the night sky.

5

Horses: Learning to Dance

My horse has a hoof of striped agate;
his fetlock is like fine eagle plume;
his legs like quick lightning.
My horse has a tail like a trailing black cloud.
The little Holy Wind blows through his hair.
My horse with a mane made of short rainbows.
My horse with ears made of round corn.
My horse with eyes made of big stars.
My horse with a head made of mixed waters.
My horse with teeth made of white shell.
The long rainbow is in his mouth for a bridle
and with it I guide him.
I am wealthy because of him.

—From a Navajo song

When the horse negotiated the land bridge [from America to Siberia millennia ago] . . . he found on the other end an opportunity for varied development that is one of the bright aspects of animal history. He wandered into France and became the mighty Percheron, and into Arabia where he developed into a lovely poem of a horse, and into Africa where he became the brilliant zebra, and into Scotland, where he bred selectively to form the massive Clydesdale.

He would also journey to Spain, where his very name would become the designation for gentleman, a caballero, a man of the horse. . . . and in 1519 he would leave Spain in small adventurous ships of conquest and land in Mexico, where he would thrive and develop special characteristics fitting him to upland plains. . . . From later groups of horses brought by other Spaniards . . . a few would escape to become feral, once domesticated but now reverted to wilderness. And from these varied sources would breed the animals that would return late in history, in the year 1768, to Colorado, the land from which they had sprung . . .

—James Michener
Centennial

When I was six years old my dad bought me a horse to ride to school. We were two and a half miles from the two-room schoolhouse and there were no school buses. The kids who didn't have their own horses were picked up for the journey in a covered sleigh, or what we Canadians call a cutter. The cutter had a small stove in one corner, which was kept burning most of the winter. No one thought too much about the cold—forty below was normal. Only when the thermometer hit sixty below was it considered cold enough to cancel classes.

The horse my father bought me, a six-year-old liver chestnut mare, was very aptly called Trixie. She was what my mother called "a dickens," and had a million clever methods for getting her own way. I particularly remember one hurried morning when I was late setting off for school. All of my cousins had already left, and as I ran to the stable gulping from the cup of warm milk my mother had shoved into my mittened hand, I was hoping Trixie would help me make up for lost time. The horse, however, had other ideas.

We started off fine, with Trixie doing her best racehorse imitation, but suddenly, after a quarter of a mile, she inexplicably stopped dead and refused to take another step. There was a brief skirmish, which ended when she bucked me off onto my head and then trotted away down the road toward her nice warm stall.

I picked myself up and stomped furiously home, where I immediately went to the laundry room and picked up a couple of clothespins. Trixie was going to learn a lesson. When I entered her stall she gave me her innocent look, jaws busily munching hay. Before she could even blink, I had reached up and clipped a clothespin to each of her astonished ears.

Though the "lesson" was certainly lost on her and didn't affect her in the least, it does make me laugh to think how important a part of the TTouch ear work has become since that first impromptu experiment in horse training.

Dependence on horses for transport was a rich and integral part of my childhood. I took it for granted then, but perhaps that's why today I have such a visceral feeling of gratitude for their role in human history. Horses have befriended us, amused us, and inspired us across the centuries. They are and remain the magical stuff of legends from Pegasus to Secretariat, from Alexander the Great's Bucephalus to Anna Sewell's Black Beauty. Even in an age where people know more about "horsepower" than horses, the image of the Marlboro Man, calm in a whirl of dust and galloping hooves, is still used to speak to us of freedom and mastery.

Legends and myths reveal that we intuitively honor the horse as a creature of sensitivity and intelligence. Yet at the same time, the way we generally train horses to submit to our will and authority shows that we also believe that they are slow to learn, naturally resistant, and need force to bring them under control.

It's an understandable duality: horses are large, powerful, sometimes volatile creatures, and their moves can be more than dangerous—they can be lethal. Fear of that power can easily lead to the belief that safety can only be assured by continuously enforcing dominance.

Even as a kid working in Alice Greaves's stable, my job was to go out every Friday and systematically "work over" the rental horses, giving them a licking with a rubber bat just to "keep them in line and remind them who's boss." Of course, I had to repeat this pattern regularly because the horses didn't truly learn to change their behavior, they simply learned to fear me. This meant that they would continue to try and get the best of me whenever they saw an opportunity.

I can't believe any well-balanced person really wants to beat up an animal. God knows I always felt bad doing it. Our ancient, intuitive recognition of the horse as a sensitive being is our deeper wisdom.

Certainly there are times when firmness is not only necessary, it is the only safe response to a deliberate challenge, but why struggle to control a thousand pounds of muscle when you can achieve an even better result by communicating with just a few pounds of brain?

When I was first demonstrating TTEAM in Germany, a famous German horse journalist heard that I was training horses in a new and revolutionary way—by increasing their ability to respond intelligently. She wrote me a letter that said in effect, "My dear Linda, it's impossible to say that horses have intelligence. Only humans have intelligence."

But what is intelligence? My teacher Moshe Feldenkrais used to love to paraphrase something Albert Einstein once said. "*Intelligence*," he would tell us students in his intense way, "*is measured by the ability to adapt to changing situations.*" And that's exactly what the TTEAMwork is all about—increasing, both in our animals and simultaneously in ourselves, that ability to respond to situations with fresh choices rather than inappropriate habitual behavior.

Moshe taught us that one way to accomplish this was through the careful execution of a system of subtle movements that interrupt habitual response (see page 11).

Over time and through experimentation I discovered a way to apply these principles of education to equine training with a system of ground and labyrinth exercises we named TEAM (Tellington-Jones Equine Awareness Method). By setting up different obstacles (see page 267) in a variety of patterns and then asking the horse to shorten or lengthen his step, to hold back a little bit, to step through obstacles and move his body without the control of the rider on his back, we created an effective way of breaking habitual responses. The result, we discovered, was a dramatic increase in a horse's interest, willingness, and ability to learn, as well as amazing improvements in confidence, balance, coordination, patience, and the ability to concentrate.

What we also discovered was that the ground exercises make a world of difference in the relationship and communication between horse and trainer.

Instead of an unseen master dominating him from his back, the horse experiences someone who is running and walking right along with him, connecting to him through guidance rather than domination. And the riders, too, down on the ground beside the horses, experience the animals from an entirely new perspective. The human is very definitely in control, but it's not a control in which the effort is to bend the horse's will. It's a control that comes from leading the horse the way a dancer leads a partner, in an exercise that unites horse and trainer in the exhilaration and success of a cooperative venture.

▷ Maggie

Horses appreciate respect. Being asked to "think" really stimulates them. Why, then, do we cling to the idea that horses can't think, even though many of us know better? Because of misunderstanding and miscommunication between the species. Because we seldom conceive of the horse as having a brain and nervous system that react to life with the same emotions as our own, with fear, guilt, gratitude, grief, and joy. As a result, it's hard for us to see behavior patterns like resistance, aggression, or personality problems as caused by anything other than stupidity or willful nastiness.

Take, for example, Norma's experience. I received her call, as I do so many, because she was desperate. A teacher of learning disabled children, Norma lives in a small New England town, where for many years she's kept her own horse as well as several others she boards for people.

A few months earlier, she told me, she had been given a beautiful three-quarter-Arabian filly named Maggie by a friend who had become unable to handle the mare's dangerous behavior.

Maggie, a three-year-old, was devoted to this first owner, but as a yearling had become a problem. She was aggressive, impossible to lead, and totally unwilling to have her head touched. She had great trouble paying attention and often exploded at the smallest distraction. Because she was lacking in confidence, she'd crowd up against the person who was leading her, a common trait in nervous horses.

According to Norma, whenever the vet, the farrier, or even a person who smelled in any way medicinal, arrived on the scene, the mare would "go crazy." She'd try to attack the person, snaking her head out over the door of her stall with her ears pinned back. If anyone tried to enter the stall, she'd plunge around wildly, kicking at the walls.

But Norma loved her. "She's so intelligent and sensitive," she told me. "She can tell what someone is going to do even before they do it. The solutions people are suggesting are making me heartsick."

One local trainer told her that the best method of breaking the mare's bad habits would be to tie her head up high and keep it that way until she was too exhausted to put up any more resistance. Another trainer's solution was to tie up the horse and then systematically beat the fight out of her.

Norma was convinced there had to be a way to avoid such drastic measures. She called several leading veterinary centers and a university famous for its expertise in animal behavior. One of the experts, after hearing her description of the problem, said that unless the mare was a valuable sports or competitive horse, trying to change her behavior would be too difficult. She advised euthanizing the animal.

"I was so insulted and shocked," Norma said. "It was like being told to put down a friend."

Finally, Norma spoke to a behavior specialist at a celebrated national center for equine studies. He gave Norma four alternatives: she could starve the horse so she would be too weak to be aggressive, she could put an electric shock collar on her, she could tie the mare into her stall and flood her with frightening stimuli until she no longer reacted, or she could try TTEAM.

As with dogs, trainers resort to these stringent remedies for aggression not because they are cruel people but because these traditional methods are the only choices they know to be available. Gradually, however, new perceptions of animals are coming to the fore. *Newsweek* magazine in a cover story entitled "The Wisdom of Animals" (May 23, 1988) reports that:

> Creatures as different as pigeons and primates are dazzling scientists with their capacity for thought. Comparative psychologists have gone from wondering whether apes can comprehend symbols to detailing the ways in which they acquire and use them. Other scientists are documenting similar abilities in sea mammals. Still others are finding that birds can form abstract concepts. The news isn't just that animals can master many of the tasks experimenters design for them, however. There's a growing sense that

many creatures—from free-ranging monkeys to domestic dogs—know things on their own that are as interesting as anything we can teach them.

The veterinarian at the equine center Norma called had read about TTEAM, and while she had not seen the method in action, she told Norma that she did believe it might be the best approach for an aggressive and obviously fearful horse.

Norma was delighted that someone had understood that the filly's problem was fear. She had worked with enough disturbed kids to recognize the interplay between panic and aggression, and she brought it up when she called me to discuss the mare's history.

Maggie was orphaned three days after her birth, and had been born so sick with septicemia (blood poisoning) that she was unable to stand. Her navel was infected and had to be cauterized, and her joints were so stiff and swollen they had to be manually forced into motion. Five times a day the vet came and gave the horse painful injections.

"She was just a little bag of bones with feeding tubes up through her nose to her stomach because she was too weak to eat," Norma said. "It all went on for so long that she developed bedsores from lying down."

Miraculously, the little filly not only survived but grew strong. As Maggie approached her first birthday, however, the buried terror and emotional trauma of those first life experiences began to surface, resulting in her erratic and explosive behavior. She was like a smart juvenile delinquent who turns bad because she can't trust anything.

What TTEAM could accomplish here, I thought, was to release Maggie's old fears, which would allow her to change her habitual reactions. I suggested that Norma begin by slowly introducing Maggie to the TTouch and the ground exercises.

At the start, Norma needed to concentrate particularly on stroking Maggie with the wand, because the stroking keeps animals focused and helps them maintain their self-control. Also, it allows a horse to fully sense its own body.

People who do things like sit at desks day after day lose touch with their bodies. In the same way, horses who live confined in stalls where they can't roll or rub up against trees or interact physically with other horses can lose awareness of parts of their physical selves. Many so-called problem horses for instance, are not aware of their own hindquarters. Stroking with the wand expands their sensory image of themselves, which in turn promotes self-confidence.

I then suggested that in addition to working with Maggie herself, Norma do so with a friend as well, in order to break down the mare's fear of strangers. This person would first win her trust with the TTouch, and then lead her through ground exercises like the Dingo and the Taming of the Tiger (pages 271 and 273).

I suggested the Dingo as a preparation for the Taming of the Tiger. The Dingo teaches horses to come forward with trust and ease when asked, and not to pull back. With the Taming of the Tiger the horse learns to calmly accept restrictions like being tied instead of flying into a panic. We use this combination of ground exercises for dogs, too, when they are fearful and aggressive with groomers or veterinarians.

These exercises were extremely helpful to Maggie, and the wand became a key tool. One day, for example, Norma looked out of the barn door to see Maggie out in the field lying on her side, her legs thrashing violently. She had become entangled in some loose wire and her panicky struggles were only making things worse. Norma immediately grabbed a wand and ran out to her. Even the sight of the wand seemed to lessen Maggie's fear, and when Norma stroked her and toned to her, the mare relaxed enough for Norma to remove the wire with no further trouble.

When I heard from Norma again with the latest update on Maggie's progress, the news was mixed: while the mare still reacted fearfully to visits from the vet, many of her other neurotic symptoms had changed dramatically. When the horse had first come to live with Norma, she had refused to leave the barn to graze in the field, and even when she was led outside, she remained huddled against the barn, trembling. She fought any approach to her head and was so terrified of water that the smallest puddle could send her into a fit.

"Now," Norma says, "she can't wait to run out into the field. She comes happily when called, wades through water to her knees, and is a breeze to bridle and to lead."

It may not be possible to solve all the problems a horse has, but

Maggie and Norma have come a long way and have found a common language with which to continue the work together.

One of the significant ways in which Norma was able to help Maggie was by following a program of TTouch called "Getting Started with TTEAM" (see page 108). This step-by-step guide was organized by my sister Robyn.

Robyn is as inseparable from the development of TTEAM as bark is from a growing tree. The second-to-youngest in our family of six, Robyn started coming to California to spend summers at my school for horsemanship when she was eleven. I can still see her, a shy adolescent with serious blue eyes and an impressive talent with horses. Then one year the shy and dedicated student suddenly metamorphosed into an ebullient and gifted assistant instructor and, to my delight, an assistant manager for the school as well.

Today, Robyn is like three people: one is the mother of two children who, at ages six and nine, are the latest members of the family to join the tradition of "riding Hoods." Another Robyn has a wild streak as adventurous as her namesake; in partnership with her husband, Phillip Pretty, and our sister Susan, she began importing, breeding, and selling Icelandic horses. Icelandics, the horses of the Vikings, are fuzzy, small, yet wonderfully sturdy horses who move with such incredible speed and sureness of foot that I always think of them as the sports cars of the horse world. Twice yearly, Robyn takes off on horse-buying safaris to the untamed tundra at the top of the world.

And finally, and very importantly to me, she is a brilliantly innovative and perceptive teacher and interpreter of TTEAM, who somehow manages to edit and produce our quarterly newsletter, and see that it goes out to over four thousand subscribers.

Robyn brings something priceless to TTEAM: where I approach my work intuitively, often finding it difficult to verbalize what it is that I am actually doing, Robyn analyzes and organizes what she sees. Frequently, in talks following a workshop taught by both Robyn and myself, she will comment to me about some new and interesting thing that I did.

"I did?" I'll say to her. "Really? What was it?" And wonder of wonders, Robyn will be able to reflect my work back to me in words, while also adding the benefit of her own humor, wisdom, and experience.

HIGHLIGHTS FROM "GETTING STARTED WITH TTEAM"*

If your horse is head shy, difficult to catch, or untrusting, you may see a real difference in just a few sessions of using the Clouded Leopard TTouch. Hold the horse with your left hand on the left side of the noseband. With your right hand, make small, single, random, clockwise circles on the forehead, around the base of the ears, and down the face. Five or ten minutes of this TTouch can help a nervous horse stay more focused, and can result in a wonderful rapport between you.

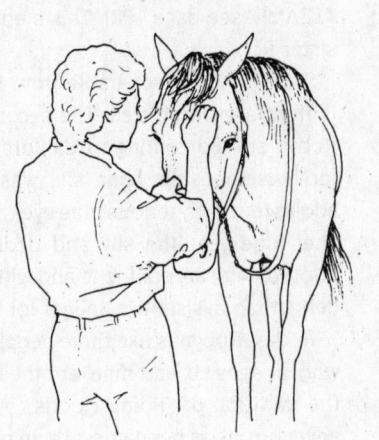

Firmly manipulating the edges of the nostrils develops tolerance and prepares your horse for tubing by your veterinarian, if necessary. Surprisingly, horses really love this movement.

* A monograph of the entire program, "Getting Started with TTEAM," can be ordered from the TTEAM office. See page 277.

Mouth work: To free the upper lip, steady the horse's head with one hand on the halter and insert the other hand over his top gum, just under his lip. Rub back and forth. This movement is essential for horses who are resistant, aggressive, or timid.

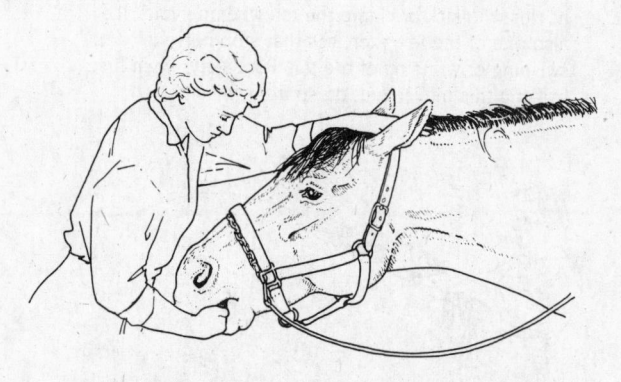

Encouraging a horse to extend the head and lengthen the neck produces a release all the way into the back, stimulates circulation in the heavy muscles of the neck, and develops trust. With your right hand held around the lower gum as shown, or cupped under the chin, slowly bring the horse's nose toward you, while with your left hand softly and without pressure encourage the extension.

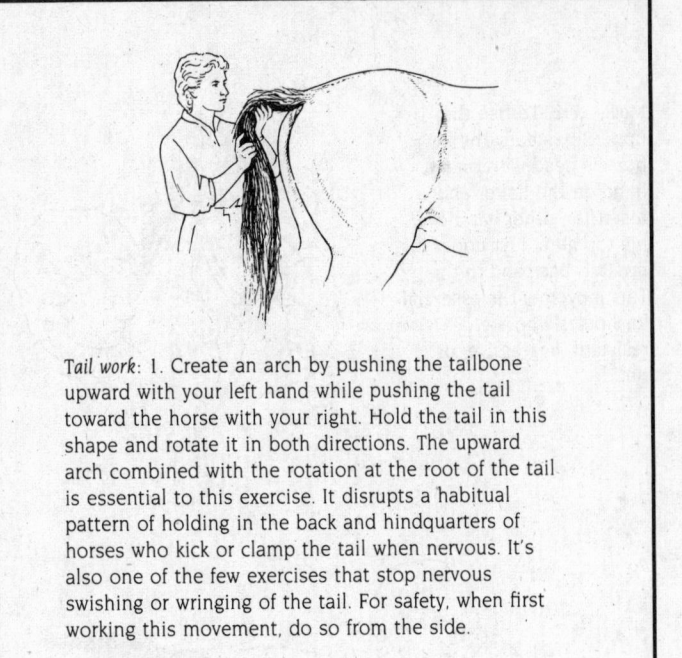

Tail work: 1. Create an arch by pushing the tailbone upward with your left hand while pushing the tail toward the horse with your right. Hold the tail in this shape and rotate it in both directions. The upward arch combined with the rotation at the root of the tail is essential to this exercise. It disrupts a habitual pattern of holding in the back and hindquarters of horses who kick or clamp the tail when nervous. It's also one of the few exercises that stop nervous swishing or wringing of the tail. For safety, when first working this movement, do so from the side.

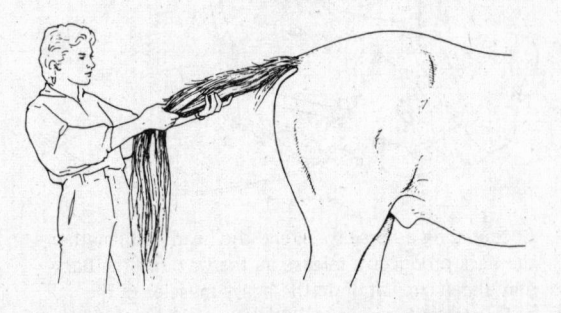

Tail work: 2. Slide your hands down the tail and pull, then hold for approximately six seconds, and release very slowly. A fast release is counterproductive to this movement. This tail pull seems to open the spine, often causing a horse to take a deep breath.

GOING BEYOND INSTINCT: THE FOUR Fs

For me, one of the most interesting things that Norma said about Maggie was how much the mare reminded her of the learning disabled kids with whom she worked. Quite a few of the children were highly intelligent but simply couldn't be reached through traditional teaching procedures. Repeated failure had made them fear the whole learning experience to the point where they froze up whenever confronted with it.

Yet, when they were taught with new and different step-by-step approaches geared toward instilling confidence, the children, Norma said, "took off like little birds."

I had worked with disturbed and learning disabled children myself, both in the United States and in Germany, helping them through their blocks by using hippotherapy (literally "therapy through horses"). I found that the confidence the kids were able to develop through successful relationships with horses was wonderfully reflected in other areas of their lives. One day, as I was reading a monograph on learning disabilities in children, I had one of those moments of insight that come when several things you've been thinking about suddenly mesh into a new understanding.

The paper, written by Dr. Annabelle Nelson, a noted psychologist, was concerned with the limbic portion of the brain—that part which, in the simplest of terms, governs emotion, metabolic functioning, motivation, and intuition. According to Dr. Nelson's thesis, the limbic system either enhances or inhibits learning ability depending on the emotional state of the learner. An emotion like fear cues the limbic system to shut down the body's responses, while happiness and confidence stimulate the limbic system to allow a greater range of response.

Aha, I thought. So when kids are in a classroom situation that makes them feel inadequate, they can't learn because their fear actually *physically* "freezes" them. It's the same reflex that makes a rabbit turn to stone before a predator, an actor go mute with stage fright, or a terrified horse stiffen all four legs and refuse to budge.

Over and over again in my experience with animals I find fear and pain at the root of learning or behavior problems. Besides the freeze reflex there are three other instinctive ways in which all animals (including ourselves) react to fear: running away (flight), becoming aggressive (fight), or losing consciousness (fainting).

While all creatures in the human-animal family react with what I call the Four Fs, we are particularly conscious of them in horses because of their size and sensitivity, and because of the special demands we place on them. In the wild, horses will normally flee from threat rather than stand and fight, but in their restricted domestic role, situations they experience as fearful can trigger any of the four reflexes, depending on the personality of the horse and the circumstances.

Some individuals, like Maggie for instance, turn their fear into "fight" or aggressive actions—biting, kicking, and, when pushed to the point of feeling cornered, attacking. Others, when put in a position where they can't escape being beaten or anything else they perceive as a threat, either lock all four legs and refuse to move or lie down and give up. I've actually seen horses that are so terrified of the loading van that they collapse on the ramp rather than be forced inside.

As soon as we realize that what we label as aggression is usually reactive and not necessarily intentional, we ourselves stop counter-reacting with our own primal responses of anger and force.

To react instinctively is to stay within an unconscious, repetitive behavioral loop. As soon as the light of awareness enters the situation, the loop is broken, action becomes conscious, and cooperation replaces confrontation. We are free to learn, to *act* instead of *react*, whether we are four legged or two, whether we bark, whinny, or talk in symbols called words. And that freedom is precisely the goal of all TTEAMwork and TTEAM training.

TEACHING YOUR HORSE TO ACT, NOT REACT:
THREE TTEAM KEYS

▷ *Lowering the Head*

Picture a horse in a pasture coming up over a hill and encountering a fluttering kite and a little knot of shouting, running kids. He snorts in panic, throws his head high, and gallops off. Or imagine the posture of a horse shying at a flapping plastic bag, pulling back and fighting at the end of a lead, or fearfully approaching a trailer. His head will automatically shoot up.

Raising the head isn't just arbitrary motion. It acts as a switch that signals the horse's adrenal glands and entire nervous system to shift into high gear in readiness for fight or flight. His breathing quickens, his sides heave, his nostrils flare. As long as the horse's head is held high, the switch is in the "on" position. His muscles remain tense, adrenaline continues to pump into his bloodstream, and he is on a chemical countdown for explosion.

The discovery that this red alert signal turns off as soon as the horse lowers his head and normalizes his breathing was of major importance to TTEAM. From virtually the first moment a horse brings his head down, his attitude changes almost completely. Where there was resistance and tension there is calmness and an entirely new level of trust. With the head lowered, the horse is ready to acknowledge the human as being in command, but he does so without fear or hostility.

I always enjoy the moment in my demonstrations when a horse finally lowers his head—I can almost hear a sigh of relief coming from those looking on. That's the point of mutual confidence where work can really begin.

Head lowering is often the first step in overcoming not only the fear of a threat from the outside world but also the fear that comes from tension or pain. I've found that sometimes what gets labeled "resistance" can be a horse's reaction to pain: cases like neck strain from the way in which a horse has been made to hold his head, or muscle stress from carrying weight when he is nervous and the body is unprepared for weight. Frequently, such a horse will hold his head high in a tension reaction.

Lowering his head and gaining his trust as a first step makes it easier for you to examine his body with the TTouch and discover those areas of discomfort and/or muscle tension that often underlie problem behavior.

We have discovered that lowering the head works in releasing tension and fear and changing the attitude not only of horses but of many creatures—think of a nervous dog with his head high and growling, a cat with its back arched and head pulled in. To understand this viscerally yourself, try arching your head back and holding it high, then let it fall forward and down. See what that does?

TWO TTEAM METHODS FOR LOWERING THE HEAD

Lowering the head is the first step in establishing your "command" position without force. It is also a key to quieting a nervous horse. Assume a squatting position and "invite" your horse to lower his head by stroking his chest and legs with a wand and signaling on the chain. Stroking encourages him to respond to your signals and body posture by lowering his head. When the human is lower than the horse's eye level, the horse is not threatened and will accept the signal to lower his head more readily. The head-lowered posture also enables the horse to relax and lengthen his whole body, especially the muscles through the neck, back, and pelvis.

Ask the horse to lower his head by using gentle pressure with both hands. The left hand uses a steady pressure while the right hand moves the horse's head with a gentle, give-and-take, left-to-right pressure. This exercise helps overcome the fight-or-flight reflex and encourages the horse to relax.

▷ The Mouth

I first consciously understood that working the mouth of an animal, like lowering the head, is a major key to changing his attitude and opening the way for communication in 1979. It was at a lecture for graduate Feldenkrais students at the Institute of Humanistic Psychology that I learned there is more to the mouth than tongue, teeth, and lips. The guest speaker was an Israeli neurologist specializing in the neurological basis for learning dysfunction in children.

In describing symptoms of neurological disturbance he spoke of the diagnostic importance of the mouth. Mouth and lips that are held in either abnormally slack or overly tight positions very often indicate a limbic dysfunction (the limbic system governs the emotions). He did not, however, indicate that there was anything to be done about it. I noted this information with interest and filed it away in the back of my mind, but it was only several months later in Germany that I understood the enormous value of what I had learned and how to use that knowledge.

I was in one of those immaculate German farmyards that are all neat brickwork and flowering borders, examining a small bay jumper. The mare stood in front of me gazing off into the distance. She was withdrawn and disinterested. Her attitude of total indifference was frustrating, but of course that's why I was there. Discouraged by her unfriendly and aloof disposition, Danzig's owners had called me in on a consultation. "She just doesn't care about people, won't have anything to do with anyone," they complained.

As I began my explorations of Danzig's body, I suddenly recalled the words of the Israeli doctor. The mare's mouth was extremely stiff and tight, the lips compressed. Well, I thought, if the mouth is an indication of limbic disorder in humans, why not in horses? Why not try to affect Danzig's attitude by working her mouth?

For the next four days I concentrated only on the mare's mouth and lips and did not work the rest of her body. (For details on how-to, see page 109.) By the fourth day we were all cheering. Her disposition had altered radically. The mare was actually pricking her ears up eagerly whenever someone approached her stall. She no longer avoided eye contact and her gaze was friendly and alert.

We never did learn the causes for Danzig's withdrawn attitude, but obviously she was feeling emotionally out of balance. People who live closely with pets know from experience that emotions can run as deep in animals as they do in ourselves, and even hardened skeptics are moved by tales of animals who travel long distances to rejoin beloved owners, who die grieving for absent masters, and who come to the rescue of humans in mortal danger.

Animals actually do have a wide range of emotion: take homesickness, for example. Over the years I've observed that many horses become severely depressed when they are purchased and shipped to new homes and sometimes even to strange, new countries. They miss the language they are used to hearing and pine for their former environment and all their human and animal companions.

I remember the case of Peroschka, a twelve-year-old black Hungarian mare at the equestrian training stable of Ulla von Tersch-Kaiser in southern Germany. Ulla, an old friend, called and asked me to have a look at the mare because she had just bought her and was very disappointed in her personality—she found the horse to be singularly unfriendly and unresponsive.

But when I saw the mare and noted the droopy way she stood and the listless look in her eye, it seemed to me we were dealing more with a case of acute homesickness than with a naturally cold character. So we worked on Peroschka's mouth and body with such gentle attention that she began to wake up to us, to feel a new connection to these humans she had eyed as strangers. Just as importantly, we changed her environment so that she would feel less lonely. Ulla's young riding students came in the afternoons after school and sat in the stall and talked to her and groomed her, and gave her the time and affectionate care that she needed. Peroschka rewarded us by becoming one of the friendliest horses in the whole stable, everyone's favorite for the next eight years.

There's a postscript: Ten years later in 1989 when I was again visiting Ulla, I asked about the mare and was told she had become stiff, arthritic, and blind in one eye, and had been put out to pasture a few miles away. When Ulla went to visit her, the horse was very depressed. Seeing her like this, Ulla brought the mare back home immediately and the last I heard she was quite perky and very happy to be ridden and appreciated once more.

It seems perfectly natural to imagine that nothing could be nicer for old horses than retirement, but they can feel isolated and lose their zest for life in much the same way as humans can when they are forced to retire or are placed in nursing homes. Like Peroschka, older horses are often much happier when they are given light work and a chance to stay in familiar surroundings with friends.

As with other animals discussed in previous chapters, TTouch mouth work is invaluable for horses who have problems with nipping, biting, and aggressive or abnormal behavior. If your horse has any of these problems and/or can't stand a direct TTouch on his mouth, try administering the mouth work through a wet towel. In our experience with these types of horses, mouth work is essential if you want a lasting change in behavior, so you just have to get in there and work on the mouth, lips, and gums. More than likely, you'll see very satisfying and dramatic results.

Working the nostrils, lips, mouth, and jowls of horses also helps to prepare them to be cooperative for dental work, tubing, deworming, and bridling.

▷ Ears

Intuitive knowledge, it seems, can flow along for years like an underground stream, rising ever closer to the surface until one day, when the time is right, it just comes bubbling up into the light of consciousness. Ear work, which today plays a major role in the TTouch for all species, had just such a gradual development.

When I was twenty years old and newly married, my husband, Went, was hired to teach English and math at a boarding school in Rolling Hills, California. Chadwick was a school for the sons and daughters of privilege and the children of movie stars like Judy Garland, Dean Martin, and Bette Davis. There were lawns as green as an Irish shamrock, tennis courts, and a stable where I usually had four to six horses in training.

Several days after our arrival the headmistress called me into her appropriately book-lined office and spent an hour politely but firmly grilling me about my life and character. Satisfied, she then appointed me dorm mother for the senior girls and eighth-grade social studies teacher.

Teaching has always come naturally to me—I became a riding instructor at the age of thirteen—but dorm mother? How was I going to "mother" girls who were only a few years younger than I was?

As it turned out, the girls accepted me in the role of "dorm older sister," and I had a wonderful if somewhat grueling time staying one jump ahead of my students in the social studies textbook while also taking charge of the equine program and the stable.

I used to get up at dawn and go down to the barn to feed and water the horses. One morning I came in to find that Bint Gulida, a top endurance racehorse and my favorite mare, was very sick. The vet confirmed that it was colic, an impaction or intestinal blockage, which meant that she was no longer able to defecate. Horses have an unusually long intestinal tract with many bends and folds, which makes such blockage a common occurrence.

The vet oiled her stomach through a nose tube, and injected a muscle relaxant to move the impaction, but two days later the blockage still remained. In 1960, surgical knowledge was not as advanced as it is today and horses mostly died, so the vet regretfully recommended we put poor Bint Gulida down.

The mare's temperature was subnormal, which is very serious in horses. Her pulse rate was a high eighty (thirty-two to forty is normal). I understood the vet's choice, but there was no way I was going to give up without one last fight for her.

By the time I reached my decision it was late and everyone had gone to bed. I walked the horse across the dark and silent campus and into the dormitory laundry room, which was warm and dry and had a cement floor. Gulida was in such intense pain that she didn't even look around and wonder what we were doing in this unusual place.

At this point the TTEAM system was yet to be discovered, but I was receiving the first glimmerings. I began to work on the horse entirely intuitively. Gulida's temperature was ninety-eight, very low for a horse, and her ears were damp and clammy. In desperation, I wrapped her back and bloated belly in wool blankets soaked in very hot water. With both my hands, I rhythmically lifted her belly, pushing up toward her flank, holding and then releasing, and for forty-five minutes I worked on her ears, stroking from base to tip over and over and working them between my fingers until they felt dry and warm. Hours went by and then, unbelievably, the impaction passed. I led Gulida back to her stall just as the dawn mist was rising from the lawn and the first birds were starting to chirp.

Twenty years later, although manipulation of the ears had become an important aspect of the nonhabitual TTouch body work we had developed for horses, I still had not yet made a conscious connection between ears and healing. At that time I was demonstrating the techniques before large audiences at German equine trade fairs. It was at such a fair that a crucial piece of information came to my attention.

Dr. Christina Kruger, a veterinarian, visited me after one of my demonstrations. "I'm curious," said this competent looking stranger in a neat skirt and cashmere sweater. "Where did you study acupuncture? With Dr. Brady Young?"

"No, I don't know acupuncture," I replied, mystified. "Why are you asking?"

According to Dr. Kruger, although I was completely unaware of it, I was working with meridians, a system of interconnected energy points that are used by acupuncturists to balance and heal the body. Christina and I began talking and when it was clear we had much information to share, she invited me to stay at her home and introduced me to the theory and practice of equine acupuncture.

In recent years, acupuncture for both humans and animals has been increasingly accepted as effective by traditional medical practitioners. The ears, I was fascinated to learn, are of major importance because they represent a miniaturized version of the entire body and its organs and systemic processes (reproductive, digestive, respiratory, endocrine, and circulatory). Various points on the ear can be used to stimulate different responses. The tip of the ear, for example, is the point used as an antidote for shock, while at the base of the ear are points to stimulate the reproductive system.

I spent the next few months staring at a small, rubber Chinese horse with the acupuncture points marked on it and steeped myself in textbooks by experts on the subject. And later, back home again in California, the healing value of ear work and its practical application became clear to me.

I was visiting a breeder, taking a tour of her barn, when we discovered one of her mares lying down in her stall, suffering with severe colic:

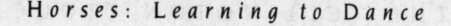

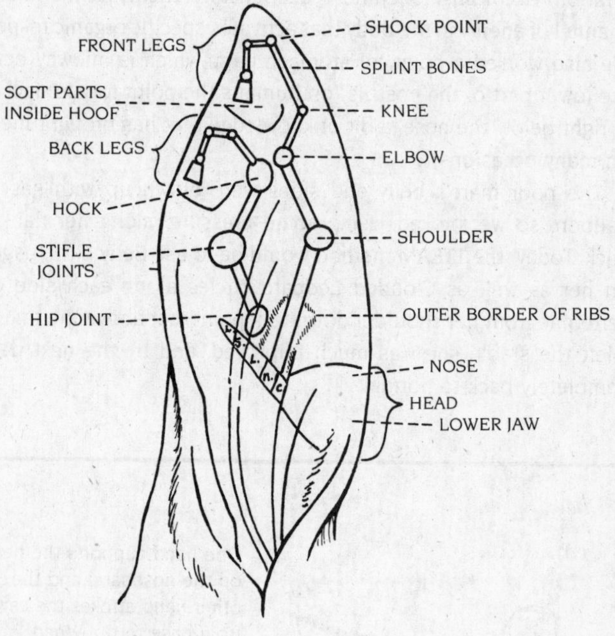

FRONT LEGS

SOFT PARTS
INSIDE HOOF

BACK LEGS

HOCK

STIFLE
JOINTS

HIP JOINT

SHOCK POINT

SPLINT BONES

KNEE

ELBOW

SHOULDER

OUTER BORDER OF RIBS

NOSE

HEAD
LOWER JAW

EAR WORK

While it is helpful and interesting to note these
acupuncture points, knowing their exact locations is
not necessary when doing the TTouch. Stroking and
circling on the ear help a sick animal to recover and
keep a healthy animal fit.

the horse's respiration was one hundred—fourteen is normal. It was a Sunday afternoon and there was no vet available. There's nothing for it, I thought, but to go to work on her ourselves.

We stroked the ears, pressing the ear tips for shock, and pressed all along the triple heater meridian, which affects digestion and respiration. According to Chinese acupuncture theory, a meridian is a channel of energy in the body that activates specific organic responses. We also worked a point for stomach pain, which is midway between the lower part of the nostrils (on humans the point for stomachaches is right below the nose, a bit of knowledge that has brought me relief on many occasions).

The poor mare's belly and sides were blown up with gas like a balloon, so we stroked, using some pressure, along her flanks and back. Today, the TTEAM method would be to use Belly Lifts (page 258) on her as well as Clouded Leopard circles along each side of her vertebrae from her shoulder down to a point just above the tail. When I left the stable she was much improved, and by the next day was completely back to normal.

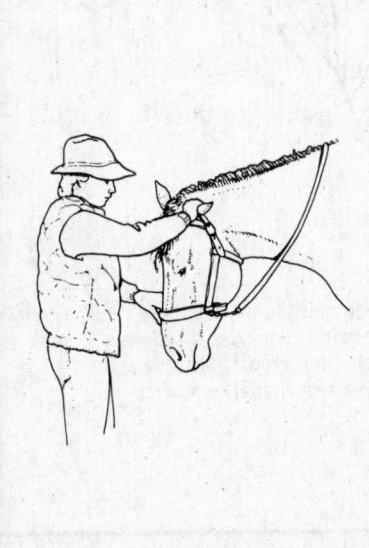

One hand supports the head on the noseband and the other hand strokes the ear from base to tip. When working to counteract shock or colic, be sure to touch the very tip of the ear. Ear work is an effective aid for horses who are overstressed, fatigued, or fearful, and it is also very successful in bringing an animal out of shock or preventing shock from occurring. Ear work can actually make the difference between life and death.

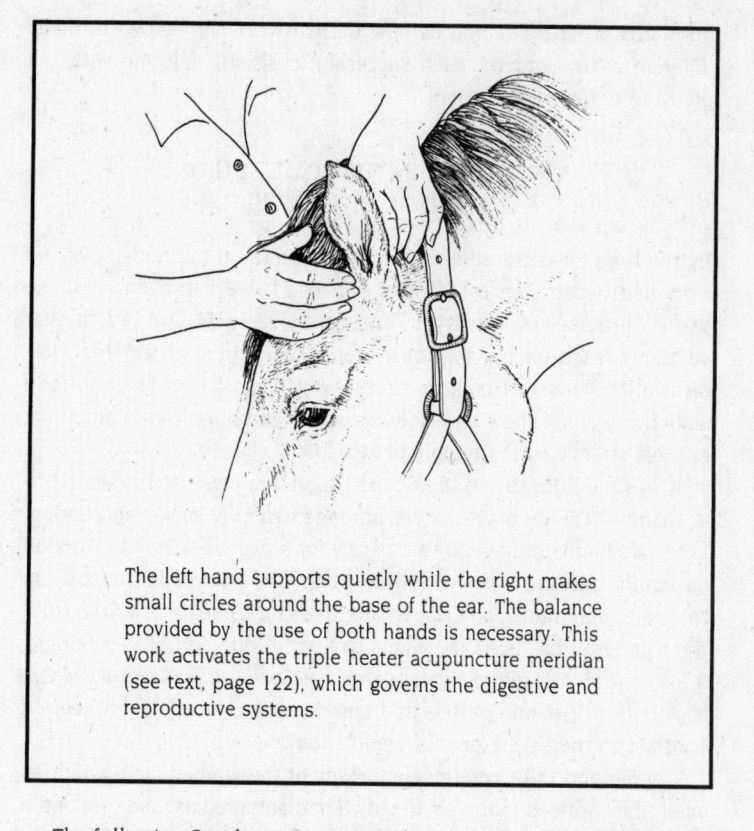

The left hand supports quietly while the right makes small circles around the base of the ear. The balance provided by the use of both hands is necessary. This work activates the triple heater acupuncture meridian (see text, page 122), which governs the digestive and reproductive systems.

The following Sunday at an equine clinic, I demonstrated the TTEAM ear work, showing the group how to move the ears in a circle, flop them back and forth, and how to stroke the ears in order to work all the acupuncture points. I still wasn't really sure whether the TTouch work was responsible for the recovery of the previous Sunday, but I told the story anyway.

"Of course it's understood that you always call a vet for colic," I said, "but in case you can't get to one or even while you're waiting, there's no harm in working the ears and doing the Belly Lifts. 'It may help and it can't hurt,' as my grandfather used to say about his remedies."

It's been a long road from Edmonton and my horse Trixie with the clothespinned ears. At the TTEAM office we now receive literally hundreds of letters telling us how the ear work has helped in many different cases, and yet each successful case still fills me with the pleasure of fresh amazement.

ADVENTURES IN THE SOVIET UNION:
A BRIDGE OF HORSES

By the time I was traveling to the Soviet Union in the mid-1980s, ear work had become an established part of TT. We had even begun to use it on ourselves. It was not unusual to see a TTEAM practitioner suddenly reach for her own or a friend's ear and start working away on it at the onset of pain or stress—which must have been an odd sight if you didn't know what she was doing. Certainly, it was something entirely strange to the people of Moscow.

It was my third trip to the Soviet Union and I was accompanied by a "team" of TTEAM practitioners, among them Ellie Jensen and Copper Love. We had been invited to Moscow for a two-week stay to work on an equine research project with a group of Soviet veterinarians. On this particular afternoon Copper and Ellie had come to visit St. Basil's, the wonderful basilica that seems to float at one end of Red Square. It always reminds me of a spun-candy castle out of A Thousand and One Nights, its stripes and pastels and onion-shaped domes a lighthearted contrast to the severe gray of Lenin's tomb.

Copper and Ellie were inside looking at the stained-glass windows when they noticed that a small crowd had gathered around one of the pews. A man was lying there ashen-faced, his breathing labored and eyes glazed, his wife crying and distraught beside him. Ellie and Copper went over and asked through their interpreter if they could help. Oh yes, please, they were told. And so they sat down with the man. Copper began working his ears with particular emphasis on the shock point while Ellie worked on his feet. Meanwhile, someone rushed off to call an ambulance.

Within three minutes the color had returned to his face and his breathing had stabilized.

"At first the crowd was quiet and you could tell they thought we were pretty bizarre," Ellie told me. "But by the time the ambulance arrived they were very interested. And the wife kept grabbing our hands and shaking them—it was great. I don't know what happened later, but at least the man didn't die waiting for help."

Over the years we've documented a great many similar instances, cases where lives were saved because someone knew how to apply ear work in the first crucial moments before professional help was available. Of course, we don't believe it should replace other first-aid techniques, but many TTEAM members who are also nurses and paramedics report that it has been a very successful addition to their regular emergency procedures.

Our project with the Soviet veterinarians was extraordinarily exciting. It was conducted at the giant Bitsa Sports Complex on the outskirts of Moscow, site of the 1980 Olympics. The purpose of the research was to measure the effect of the TTEAM methods on adrenal function and stress levels in sport horses.

Bitsa's chief veterinarian, Dr. Nina Kanzina, had arranged for eight veterinarians to take two weeks off from their usual assignments to have the opportunity to work with us.

Nina is solid and strong both physically and in character. She cares tremendously about animals—there were three or four horses at Bitsa who could no longer be ridden, but rather than put them down, Nina made sure they continued to be loved and cared for. I was fascinated to learn that not only does she oversee the care of Bitsa's one hundred and fifty horses, but she also has the mandate to change a trainer's program if she thinks it is not good for the health of the horse. I had never heard of such wide-ranging authority for a veterinarian. She never abuses this authority; she is very clearly an advocate of all the horses at Bitsa.

The Bitsa Complex is a spectacular government-run riding center where horses and riders are trained for national and international equine competition. There are two arenas, one for warm-up and the other a huge show hippodrome with two thousand seats and a fifteen-foot stained-glass window at one end. The space is so vast and airy that it easily holds twenty-five to thirty horses working there at one time, both at jumping and dressage. Beneath the arenas are stables housing the one hundred fifty horses in training for jumping, dressage, and three-day eventing.

Sometimes I would arrive in time to catch the early-morning exercises of the Olympic-team jumpers, and would enjoy the heart-lifting moment when horse and rider sailed as one over the barricades. Then at nine o'clock the dressage riders would arrive to study with their coaches. They came, men and women, from every corner of the fifteen Soviet Republics, and as I looked out at them working with their horses, I thought of my grandfather and the tales of Russia he had spun for me; now I, too, had become part of the story.

Sometimes I'd have a chance to trade new experiences with the Soviets. I'd demonstrate American training techniques for them while I rode fabled horses I had only heard of but never seen before, breeds with evocative, romantic names like Bidjoni, Don, and especially Akel Teke, the golden horses of the Asian Steppes whose ancestors had been the treasure of Alexander the Great.

The Soviet veterinarians were a delight to work with. They had a love for and ease with animals that seems to be part of the Soviet personality. To test whether TTEAM brought about changes in the level of stress hormones in the blood, we used twenty horses; ten worked with TTEAM and ten did not. The horses were all top contenders who performed strenuous competitive tasks either as open jumpers or three-day event horses and they all worked out for one and a half to two hours every day.

One of the most interesting cases I was given was that of Bedouin, a six-year-old Don stallion a little over sixteen hands, a beautiful, bright-chestnut hunter. Bedouin had serious problems and had thus been taken out of training. His circulation was very poor and he had an irregular heart rate and constantly swollen legs, all interrelated conditions. He had been on medication for a year and was very depressed. The fact that his heart was monitored with an EKG machine might strike some people as unusual, but for the Soviets it's part of a veterinarian's job. Soviets take the health of their animals very seriously: at Bitsa there is one vet for every thirty horses, and looking after these thirty is his or her only responsibility.

During my exploration of Bedouin's body he was quite calm, almost lethargic, but as soon as I tried to touch his ears he fired up. He was always like that about his ears, I was told, touchy and hard to bridle. It took twenty minutes to get him to lower his head so that he would have the confidence to allow me to work on his ears. When I finally did get to them, I was amazed.

The top half of each ear was soft and flexible as is normal, but the bottom half was hard and stiff and slightly swollen, which to me indicated a circulation problem that was possibly related to the irregular heartbeat. I therefore went over both ears very thoroughly, hoping to improve the circulation, going over every centimeter with small circles until the stiffness softened and disappeared.

When I returned to Moscow three months later we reexamined Bedouin. At my suggestion, several veterinarians had been faithfully working on him daily, carefully stroking and working his ears from the base up to the tip. Also, they did the TTouch on his windpipe, which became much more flexible, as well as on his flanks, legs, and tail. During that time they saw a significant change in the horse. His pulse, eighty-eight the first day, had lowered to a normal thirty-six and the swelling in his legs had almost completely vanished.

To track the effect of the TTEAMwork, one of the doctors had continued to monitor Bedouin's heart rate with an EKG machine. He reported a measurable improvement in the regularity of the heartbeat and we were all tremendously cheered—especially Bedouin who, after three months of TTEAM was well enough to return to the arena where all the action was.

Along with Bedouin, the veterinarians had also been monitoring the other horses in our experiment and blood tests found a marked lowering in the stress hormones of the horses worked with TTEAM.

Soviet people have a relaxed way of doing things and I cemented many friendships at Bitsa during our extended morning and afternoon tea breaks. The tea room was in the veterinary center, adjacent to the arena, and we'd gather there to drink the strong, smoky brew dispensed from the samovar and sample the cakes and sweets and various Russian delectables that people brought in to share with each other. There were home-pickled tomatoes and cucumbers, smoked fish, sweet-tasting boiled potatoes, aromatic cheeses, and, best of all, freshly baked sour rye bread spread with sweet butter just off the farm.

A dignified gray cat with four white paws and a white chest presided over the room, waiting tactfully but with unblinking attention for her rightful due, a morsel here, a taste there. We became very good friends and on my last day she was given to me as a present; because I could not take her with me, she became my Moscow cat, in the hope that I would return often.

Though I treasure many memories of my travels throughout the USSR, a few images stand out—an old woman washing a pile of freshly shorn wool in a stream of water running from a pipe into the street; large herds of cattle standing in a river, above them a looming portrait of Lenin scratched on the side of a mountain; three young men in a weedy field on the outskirts of a smoggy town, riding horses with such electrifying grace they looked like centaurs.

One day I went walking up a mountain and encountered an old herder resting beside the path, his one-eyed horse saddled and tethered beside him. I stopped to chat about his horse, using my meager Russian vocabulary plus a lot of hand gestures and smiles. The horse didn't want to be touched at first, and told me so, with laid-back ears and a shake of his head. The animal soon realized, however, that this was something both different and enjoyable, and laid his scraggy old head on my shoulder while I worked on his neck.

The old man was fascinated, so I showed him how to work on the horse's ears, and on his own ears, too, in case of sickness, and we talked, with much laughter and elaborate pantomime, about how this same movement could also be good for pigs, or cows, or grandchildren.

My meeting with the old man was one of dozens of memorable encounters brought about by animals. I was introduced to the people of the Soviet Union by connecting with their fish, birds, dogs, cats, donkeys, and horses. At the Moscow Zoo, the director, Dr. Spitzen, visited with me and then took me to meet two handsome, whiskery walruses, a tumble of cheetah kittens, a monkey, and a bear.

I recently found out that TTEAM is now being used by horse trainers throughout the Soviet Union, even as far as the steppes of Asia, on my beloved Akel Tekes with their burnished, golden bodies and thick, black manes. It moves me to see the communication methods of TTEAM working on so many levels in so many places around the world, but for me, what continues to lie at the heart of the matter is the relationship between a single, individual human and a single, individual animal. It's in those breakthrough moments between a handler and a horse, a person and a dog, a child and a cat—those moments when they recognize their common ground—that the work comes alive and continues to grow.

6

Feathered Brethren

It took the whole of Creation
To produce my foot, each feather:
Now I hold creation in my foot.

—From "Hawk Roosting"
Ted Hughes

Hope is the thing with feathers
That perches in the soul,
and sings the tune without the words,
and never stops at all . . .

—From "Hope is the thing with feathers"
Emily Dickinson

FEATHER POWER

Over the years I've collected many stories from many places about the importance of animals in human life and one of the most touching was told to me by a man named Jack Snow. When his wife left him without warning after many years of marriage he was completely devastated. But every day when he came home to an empty house, his dove would fly over to greet him. She'd stay perched on his shoulder, and when he sat down to read the evening news she'd play a game, reaching down and pecking at the paper.

"She must have sensed how depressed I was," Jack told me, "and having her there gave me a reason to keep going."

It's becoming more and more evident that the friendship of animals truly helps humans young and old to heal grief, illness, and trauma. Studies show that the presence of animals can bring about dramatically positive changes in cases of severe depression and autism as well as in metabolic, heart, and blood pressure rates. In nursing homes, hospitals, and mental institutions, visiting dogs, cats, birds, and gerbils are an increasingly common sight, and staff members say that one of the biggest gifts the animals bring is a sense of hope.

It seems to me that of all the animals, birds speak to us most eloquently of hope. They soar and sing high above us; at dawn their voices celebrate the promise of a new day. The phoenix born of its own ashes, the white dove of Noah that brought the ark its first green message of a renewed world, the dove of peace hoped for around the world, the bluebird of happiness—these are the images birds spark in us, reflecting our own longings and aspirations.

Birds represent both strength and fragility, a duality that can be seen even in the very structure of their feathers. Tough enough to withstand wind and water, a feather is also supple and sensitive enough to deliver the most delicate caress, making it, with wonderful appropriateness, the perfect tool for stroking birds themselves. Holding a feather you can follow the fragile contours of a bird's body with just the right amount of flexibility and subtle pressure.

If you've ever attempted to quiet or tame a bird, you know how hard it is to approach the job with bare hands. However, where hands are threatening, a long feather reaches out and establishes contact without menace. In the TTouch for birds, feathers are used as an extension of the hand and arm in just the same way that wands are used with other animals.

Of course, even in a confined space, a frightened bird won't simply stand still and let you stroke it with a feather. That's why we often begin a session with two feathers, one on either side. With any animal, including humans, whenever the body is touched in two places at once the brain cannot react with its habitual response.

I use whatever feathers are handy—vulture feathers, turkey feathers, macaw feathers—but I'm especially fond of using a white owl feather (about twelve inches long) and another one of equal length that I try to match to the bird's own color. For smaller birds, of course, smaller feathers are appropriate, but whatever the size, the feathers should be stiff enough for the bird to feel the contact. I begin by stroking the bird on both sides of the body. Whenever I do this in demonstrations, people are stunned to see how quickly the bird quiets down and stands still.

My beautiful owl feather is a gift I received after one such demonstration. Ewald Isenbugel, the dedicated chief veterinarian of the Zurich Zoo, had invited me to present the TTouch to a group of his students and assistants at the Zurich University Veterinary School.

I was to work on a traumatized Amazonian parrot. A bright green bird a little bigger than a dove was brought out to me. He was hysterical, and when anyone tried to reach into his cage he squawked and fluttered, defending himself by pecking at the person's hands.

Toning to him the whole time, I reached the feathers into his cage and stroked them slowly against each other in a corner of the cage away from him until he settled down a bit. Then I used them to stroke both sides of his body until, after several minutes, I was able to touch him gently. After several more minutes, in which I touched his body with tiny, light circles, he was calm enough to allow me to move my hands freely in and out of the cage. After a few more sessions, more than likely, he would willingly have perched on my hand.

Although I approach all animals with respect and give them time to accept me into their personal space, I've learned that creatures vary in how soon they will permit you to touch them, and how close they will allow you to advance. Some animals are actually quite interested in people and will allow connection in a surprisingly short time. Others, like birds, are more reluctant. A speedy metabolism can bring them to an instant panic point. Heart and breathing rates rise, which further increases their terror and can send them spiraling into hysteria. This is where the calming effect of the two feathers breaks in on the fear reaction and interrupts the spiral.

And that's exactly what I demonstrated with the green parrot. Afterward, Ewald brought me into his university study and in honor of the occasion presented me with my white owl feather. Not only was the feather beautiful, but it had special properties that singled it out from the feathers of other birds. The leading edges of an owl's primary wing feathers have comblike projections while the upper surface of each feather vane is often covered with down. These adaptations make the owl a very silent and stealthy fellow in flight. They also provide me with the pleasure of using a feather that has exceptional delicacy and strength.

"You know," Ewald told me, "what you're doing with feathers is a centuries-old training method. Back in the thirteenth and fourteenth centuries falconers used to tame their birds that way." And he went on to tell me about how the falconers used to sit up all night with several bottles of wine, a warm fire, and a generous supply of candles, drinking to pass the time while they continuously stroked the bird with a feather to keep him awake yet calm. By morning the bird, his defenses worn down, would be tame.

Well, I'm happy to say there have been *some* advances since the fourteenth century; in my experience, you can now tame a wild bird in as little as three or four fifteen-minute sessions and avoid the bottle, the candles, and the hangover. The taming process is also useful in handling birds who are already tame but injured, like Archimedes Owl.

Archimedes, a great horned owl, was another resident of the San Diego Wild Animal Park who performed in the Critter Encounters show. He was very friendly and curious, and I loved to see him effortlessly turn his head almost one hundred eighty degrees, taking in the world with his magnetic, topaz eyes.

During the three years that Archimedes had worked in the Critter show, he had developed a habit bird handlers call "baiting," or jumping off the gloved hand that held him. He would make a sudden dive into space that would leave him dangling upside down, suspended by his feet from leather thongs attached to the keeper's wrist. Because of this habitual baiting, he stressed himself to such a degree that he hurt his wings.

The owl was very tense and frightened. He was not ready to let me handle him, so I worked with him through the several stages that prepare a bird to open up and relax enough to allow the TTouch.

First stage: I introduced myself by stroking him with two feathers, one on either side of his body (see photo insert). I never start with the back. Most creatures are instinctively afraid of something that comes in from above or behind them—don't you, too, feel a rudimentary prickle in the back of your neck at the hint of unseen danger?—but birds are particularly frightened of something that is perceived as swooping down on them.

The next step is to move the stroking feathers slowly down to the chest and sides, watching the bird's response. When I did this with Archimedes he was a little nervous, so we quit for a while.

The more upset a bird is the shorter you make the session; begin with five minutes, then go to ten as he develops confidence. If the bird is very upset, take a break of two hours between sessions, otherwise you can allow a break of half an hour to an hour, or you can work on him the next day or even two days later, depending on your own schedule. With Archimedes, I went to look at a few other animals and came back to him after an hour.

During the next session we were able to go through steps one and two in minutes, so we were ready for **step three:** one hand keeps a feather steady against one side of the bird. The other hand strokes the opposite side with the other feather while the fingers inch up the shaft toward the bird's body. It's a rhythmic progression, stroke and move the fingers up, stroke and move up.

The bird remains tranquil because you are keeping a steady but light pressure with the feather on the opposite side, giving him a sense of confidence and safety. By the time the moving fingers reach the bird's body, he is calm enough for you to discard the feather and try the TTouch. Usually you can get to this stage within ten or fifteen minutes, but it does vary from case to case. With Archimedes this initial calming took the entire forty-five-minute session.

When we began the last session, Archimedes was much more relaxed right from the start. This time I stroked him down the chest and out to the sides, slipping the feathers under his wings and putting slightly more pressure under the well wing. Suddenly his whole attitude mellowed and he began lifting his wings slightly away from his body.

When you reach this stage and the bird lifts his wings for you, you can almost hear him sigh and say, "Okay, I guess I can go with this. It's safe and besides, it feels good." That's your signal to slip the training feathers up under the front of the wings, encouraging the bird to open them even more. Then you can go on, sliding your fingers up the one feather until you get to the bird's body. Discard the feather and reach with your fingers through the bird's plumage and the cloud-soft underlayer to make very small Raccoon circles (see page 244) directly on the body. The pressure is very light; usually I use a two, and for small birds, a one.

While you are doing this with one hand, you are still supporting the bird with the feather in the other hand, giving him confidence and keeping him from moving off.

As we went into this phase, Archimedes kept his wings slightly spread and bent his head while a dreamy expression slowly filled his owly eyes. While he did improve right there in front of our eyes, we would have needed several more sessions to be really effective, and I was sorry not to have the time. Since then, however, the TTouch has been documented as being very successful in helping other large birds of prey with the same "baiting" habit.

Toning to an animal as you're working is a major aid, and I didn't leave it out of my meetings with Archimedes. Toning helps to control your breathing, which slows you down and focuses you on what you're doing. Breathing rhythmically and slowly into your strokes and circles is of particular importance with birds because of their speedy and excitable natures. You need to be very serene yourself, so that as you connect with each other the bird can get in tune with you and begin to mirror your calmness.

Sometimes you can influence the mood of a session with tranquil music; sometimes birds are soothed when you echo their own songs and calls to them, soft and low, or with a breathy whistling. Starting each session with music brings a peaceful continuity to the gentling process for both you and your bird.

I've also found that the same two-feather technique works like a charm in preparing a bird to accept banding, toenail clipping, or work on arthritic feet. Gentle him first with the steps outlined above and then move to the legs. If the bird is already gentle, you can skip the first phases and go straight to the legs.

Begin by simultaneously stroking down the outside of each leg, then move to the front and back, and then the inside of both legs.

After you see the bird is comfortable with this TTouch, use a single feather on one leg, again moving down the outside, back, front, and inside surfaces. If the bird is easy and calm, you can then hold the feather in the middle and slide your fingers up until you are able to touch the leg without upsetting your feathered friend.

Now you're both ready for the tiny Raccoon circles. Make sure you go up and down the leg and right down over the toes. If you need to take your bird to the vet for toenail clipping or for any other reason, these steps are an excellent way to prepare for a trauma-free visit.

There is a special way of holding smaller birds that was taught to me by Sue Goodrich, who for many years was a senior keeper in the San Diego Wild Animal Park nursery and hospitals. We were working at the Park with a crow, trying to ease the stress that had caused him to beat against his cage with his wings and damage them. I noticed that Sue cradled him in such a reassuring hold that he didn't utter even one squawk.

Sue's hold: With your palm against the bird's back, encircle his body with your thumb on one side, fingers coming over the top of the bird and around to the other side, covering his wings at the neck. Now that you're softly holding the bird over the lower neck and top of the wings, put your other hand on his breast as a gentle support.

This is the perfect way to hold a wild bird. Like a newborn human baby wrapped in a receiving blanket, it seems to feel more secure when closely held. If you find a fledgling that's fallen from its nest, you can calm it by holding it in your hand and stroking it with a feather before doing the TTouch. When trying to help an injured bird, hold the bird in one hand and using the tip of your forefinger on the other hand, penetrate gently under the feathers to make very light circles. Sometimes, although a bird may appear dead, he's only knocked out or in shock, so don't give up until you've TTouched him for as long as ten minutes.

▷ *The Ups and Downs of Wild-Bird Rescue*

My friend and student Eleanor McCulley had an extraordinary experience with a bird who appeared to be lifeless. Eleanor, a registered nurse, lives with her husband, Duncan, in the peaceful, wooded hill country near Austin, Texas, with several horses, a large macaw, a cat, and a floating population of dogs.

Animals have a way of adopting the McCulleys. "We don't have trouble keeping our dogs *in* our yard," Eleanor says, "we have trouble keeping other dogs *out*. They just hop over the fence and the next thing we know we have a new dog."

Early on an uncharacteristically freezing Sunday morning last winter, Duncan came in from his chores, went running upstairs to his wife who was still in bed, and handed her a pathetic bundle of frozen feathers. "Eleanor, you have to fix this bird I found out by the pump house," he said.

The bird was a roadrunner, just like the cheeky cartoon character, only this one wasn't saying anything like "beep beep!" He was suffering from hypothermia and wasn't saying anything at all. For the next forty-five minutes Eleanor concentrated on making tiny circles around the bird's "ears," feather-covered orifices just behind the eyes. As with other animals, ear work often brings remarkable results in cases of shock. Sometimes, within minutes, the bird will pop up and fly right out of your hands.

But the roadrunner was too far gone to respond. Eleanor wrapped him in a towel to keep him warm, holding him in one hand and turning him from one side to the other every ten minutes to alternate the work on each ear. She stared at his emaciated chest, seeing no signs of breathing or life. His eyes remained closed, his body cold. She could find no pulse.

"It was so strange," she told me. "He looked so completely dead, but from the moment I started working on him I got the clearest feeling of communication with an animal that I'd ever known—so clear it was like a presence—the way it feels when someone is in a room with you even though you can't see them."

After forty-five minutes Eleanor stopped working on the bird. "I talked to him out loud," she told me later. "I said, 'Listen, I *know* you're not dead, but please, just open your eyes or give me some sign to let me know you're still alive so that I can keep working on you.'

"And I had no sooner said it than the bird opened his eyes, looked right at me, and then closed them again."

So Eleanor went on working on the bird and an hour later, when she could see he was breathing, she began doing small lifts on his body, tiny, minuscule movements of the skin under his wings. After another fifteen minutes he opened his eyes and lifted his head. Eleanor held him cupped against her stomach and fed him water from an eyedropper.

She and Duncan alternated working the TTouch on him through the night at hourly intervals. By morning, Eleanor was grateful to see how far he had progressed.

"His eyes were very alert," she said, "and though he was skinny and malnourished from the freeze, his feathers were fluffy and shiny. And he had such great spirit—I thought he'd live."

Eleanor and Duncan both had to work and it was impossible for them to continue on the demanding schedule that would be necessary to save the roadrunner. Sadly, they decided that they would have to turn the bird over to a couple they had heard were interested in saving and rehabilitating birds.

"So we took the bird over," Eleanor said, "and they were wonderful, kind people, but they were real pessimistic about his chances. As soon as they saw him and picked him up they said, 'Oh, no, look how malnourished he is, we can almost guarantee he's not going to live.'

"My heart sank," Eleanor said. "I showed them how he would lift his head and reach out to grasp my finger and they were surprised and commented that maybe he had more strength than they thought.

"It's not that he has strength, I told them, it's that the bird has the will to live."

She went on to explain to the people about how she had developed a relationship with the bird by using the TTouch, and how if they tried it in addition to all the other good things they were going to do, she was sure the bird would have a fighting chance.

Eleanor called to tell me the people were very interested in trying the TTouch but when she called again several days later she was in a flood of tears. The roadrunner had died. If only she had kept him, he would have lived, she told me; his death was her fault.

I've found that when an animal we care for dies, guilt is often a part of grief. Did we do the right thing, could we have done better, tried harder? we ask ourselves. But Eleanor had done the best she could.

"That bird changed my whole relationship with animals," Eleanor told me. "He had a really big influence for such a little guy."

▷ *Visualization and the Swan with the Invisible Wing*

The people to whom Eleanor brought her roadrunner were obviously extremely well intentioned, but as soon as they saw the bird they just couldn't *imagine* him alive and on the wing. His condition was so poor that they believed it would be *impossible* for him to get better.

I know everyone thinks that way sometimes. I've caught myself, too, thinking about an animal, Oh, this dog will never learn, or, No way this creature is going to live. And sometimes, it's true, they don't learn and they do die. But to me, making those statements can also be exactly like sealing yourself into a room with no doors.

Actually there is always a door—imagination. "Imagine" has the word *image* in it, and also a piece of the word *magic*. To imagine is how we create the door that leads from the impossible to the possible. The magic of it is just that we don't really know how or why it works.

There is ample evidence, however, that it does. In every sport, the practice of positive visualization or "image-ining" has become an accepted tool for success. As verified cases accumulate, the medical profession, too, has acknowledged the transformative power of visualization and its beneficial and sometimes miraculous effect on disease.

An inspiring example is Norman Cousins, for many years the editor of *The Saturday Review of Literature*. In the mid-1960s, Cousins was stricken with a life-threatening disease that paralyzed most of his body, but because of his great faith in the regenerative capacity of the human mind and body, he refused to give up, even when he was advised that chances for his recovery were slim. Following a self-prescribed regimen of massive vitamin C injections and positive emotions, he completely recovered, much to the amazement of his doctors.

When we work with an animal, whether the problem is physical, emotional, or behavioral, it's important to keep that door to the possible open so that we don't limit the outcome. That's where active, positive visualization can come into the picture.

I can't think of a better example of this than Emily the swan. I met

Emily through Martha Jordan (see page 94), a.k.a. "The Swan Lady." From their home base in Snohomish, Washington, Martha and her husband, Michael Kyte, a marine biologist, travel and work as environmental consultants to the oil industry, monitoring oil refinery sites to make sure that regulations are observed and to keep the companies up to date on environmental issues.

But Martha also has a very special passion—trumpeter swans. She heads the Snohomish Trumpeter Swan Working Group and is known and respected the world over for her expertise on these very special birds.

Trumpeters get their name from their deep, resonant, bass trumpet call. Their courtship dance is spectacular—they stand facing each other only a few feet apart, fanning their wings and bobbing their heads in an elaborate ritual. If domesticated, they even like to dance with humans.

"Swans often appear in fairy tales and myths as having magical and mystical powers," Martha says. "When I went to an international conference of biologists in England, people were comparing notes about swans and commenting on how amazingly many countries revere them and consider them to be omens of good fortune and good health, not only for the humans, but for the land."

Martha had come to Seattle to attend one of my horse training workshops and had caught my attention with her riveting tales about wild swans and her experiences banding wild geese in Alaska, herding thousands of them with the help of a Border collie she had trained for the purpose. As if that weren't enough, I also found out that her husband, Mike, is the chief provider of octopus for most of the aquariums in the United States. He brings them in from the Puget Sound, which is an ocean inlet near Snohomish.

After the clinic was over and I was packing to leave, Martha came over to speak to me. She wondered if the TTouch would help Emily, a trumpeter swan with an amputated wing, who had been sent to her from Alaska for rehabilitation.

Swans, dogs, geese, octopus, and now this—how could I resist? We packed my gear into Martha's station wagon and headed off for Snohomish.

I awoke the next day to a morning crisp as apples and to the sound of six wild swans honking for their breakfast. Emily was in a pasture behind the house, under cover because her waterproof feathers had not yet grown in to protect the exposed area where the wing had been amputated.

She looked very scrawny, but even in good health she would have looked spindly and a little awkward. At four months Emily was still very much a baby. She had been found about twenty miles from her nesting grounds at Minto Lake, Alaska, virtually dead. One wing had been completely shattered by a bullet. The biologist who found her had wanted to put her down, but his wife, a wild animal rehabilitation specialist, insisted the young cygnet could make it.

Emily saw us approaching from forty feet away and immediately began flinging herself against the back wall of her enclosure in a paroxysm of terror. Understandably, humans frightened her; one of the aims of treating her with the TTouch would be to calm and tame her so she could be handled without being hurt.

We worked on her by the swan paddock in front of the house, her breast flat on the ground, legs pulled straight out behind her to keep her from struggling. Bobby and Julie, two trumpeter swans that had been rescued young and reared in captivity, looked on with interest. Within a minute or two of starting the TTouch, Emily quieted down and relaxed.

I noticed that some of the important feathers on the injured side were not growing in correctly. A few stuck out at odd angles; if they were allowed to continue in this way, Emily would not be waterproof and would be exposed to the cold and to probable sickness. As we worked I asked Martha to picture strongly the feathers growing in straight and healthy.

Emily was a perfect case for the combination of TT circles and visual imaging. After first doing Raccoon circles with a three TTouch over the swan's entire body to get her used to us, we went on to use a very soft number one Raccoon TTouch on her remaining wing.

You begin working the sound side first in most cases of injury and amputation because the injured side continues to hold and image the fear and trauma associated with receiving the wound. By working the normal side first as gently and carefully as possible, while at the same time holding a picture in your mind of the wounded side, you can actually ease and temper the fear and contraction of the injury site without causing alarm, giving the cells a message of well-being.

After completing the work on the normal side, Martha and I moved to the site of the amputation and worked all around the area where the wing was removed using the Lying Leopard, the Clouded Leopard, and the Abalone touches and keeping in mind a clear image of the "phantom" wing. We then continued on out into the air, tracing and circling the outline of where the missing wing would have been.

Whenever a limb or part of the body is amputated, the remaining nerves react as though it is still present, creating a phantom limb sensation. Some people even experience pain in the part that is no longer there. We've found that working on these areas, as odd as it sounds, helps heal the physical as well as the emotional pain and trauma of amputation, not only for animals but for humans as well.

It was wonderful to see the change in Emily at the close of the session. Her eyes were closed, her neck flopped over, and her webbed feet, now comfortably limp, looked like little furled umbrellas.

Emily's operation, however, had caused her more than pain and panic and the faulty growth of her feathers. Without the balance of two wings she was unable to run without falling over on her side. She'd start on a lopsided takeoff and then, just as she was beginning to really zip along, she'd fall over and flip onto her back. It was painful to watch—how much more frustrating it must have been for Emily.

After I left, Martha worked for three sessions, specifically concentrating on this problem by once again combining TT and visualization, imaging the missing wing back in place and Emily running along in balance. Within five days she was dashing around the pasture almost like any other rambunctious young swan.

"Her feathers did straighten," Martha reports, "and she's much friendlier."

Four months after Emily's arrival in Snohomish, she had become queen of the roost. She took charge of the education of the trumpeters, Bobby and Julie, teaching them the proper behavior of wild swans, divulging secrets like how to use one's feet to stir up the delicious goodies available in the shallows of the pond.

Emily is presently in Ontario, Canada, in a breeding restoration program where she has become a mother. The program is part of a government-sponsored drive to breed trumpeters in captivity and then turn the offspring loose in the wild to restore the swan population decimated by man's interference.

"There are thirteen thousand trumpeters on the Pacific Coast of the United States but only two thousand in the rest of the country," Martha says.

"You see, by hunting them and spoiling their habitats, we killed most of the older generation of swans in the midcontinent population, and with them died the knowledge of their migration routes. Without the previous generation's knowledge to guide them, the young trumpeters didn't know where to migrate from their breeding grounds to survive the cold winters, and so they began to freeze to death.

"We killed the swans' knowledge by killing their elders. Just as in a human tribe, if all of the senior members are killed, the tribe's traditions are lost and gone.

"So now, as we become conscious of our true stewardship of the earth and of our responsibilities, we're trying to reseed the swan population and to restore what we stole."

Martha told me, too, of some amazing methods naturalists are coming up with to teach the trumpeters migration routes—among them imprinting the baby swans to follow boats and ultralight aircraft so that later they may be shown the way south.

When I think of Emily, I think of how she became enmeshed in our changing perceptions of nature, first as a victim of our cruel ignorance, then as a recipient of our love and care, and now as a partner in the job of restoring what we have destroyed. It's my hope that her babies will be born into a time when we truly understand how to cherish them.

7

The Warm Hearts
of the Cold-Blooded:
Relating to Reptiles

Miss [Grace] Wiley had long experience as a herpetologist
and was considered one of the world's most skilled handlers
of snakes with bad reputations. Indeed, the tougher, the
meaner, the more venomous they were, the better she liked
them. . . . People came from all over to see her handle the
snakes. . . . With one of the deadly specimens nestling
affectionately in her arms, she would show to fascinated
audiences what splendid philosophical teachers and companions snakes can really be when given an opportunity.
She would usually close her talks with the observation that
deep within its heart the snake is not a troublemaker but
a fine gentleman, and that when he strikes he does so
because someone with evil intent has invaded his domain
and cornered, frightened or hurt him.

—J. Allen Boone
Kinship with All Life

QUESTIONS OF LIFE AND DEATH

The one-room schoolhouse was red and had a bell, just as you'd
imagine a frontier schoolhouse on the American prairie in a much
more innocent era than 1950. In Alberta, Canada, time moves slowly

and with kindness. I was in the fourth grade, having made my way up from the Lilliputian desks in the first-grade corner to the place by the door with the older kids. There were twenty of us from the outlying farms and countryside. In winter we labored over our copybooks to the whoosh and pop of logs burning in the stove. In the spring we ran outside at recess to yell and run and search out the first ground-cracking lady's slippers and jacks-in-the-pulpit poking up green among the dry leaves.

During one such recess in May, a little pack of older boys came bursting out of the trees with Frank, the leader, brandishing a large garter snake. Whooping and laughing, they ran around chasing us with the snake until the smaller kids, among them my younger brother, beat a terrified retreat into the schoolhouse.

I was scared, too, but I was so mad that before I could think better of it I found myself standing in front of Frank saying, "Ooooh, what a looovely snake, Frank."

"Oh, yeah?" he said. "If you like it so much, why don't you hold it?" He waved the poor serpent right under my nose. Uh-oh, caught by my own game. There was nothing for me to do but clench my teeth and hold out my hand.

To my surprise, instead of the cold and slimy wriggle I was expecting, holding the snake was really nice. It felt beautifully smooth, like a ribbon of cool satin flowing through my fingers.

That was my first "close encounter" with reptiles. Since then I've handled and worked with many types and species, from iguanas to boa constrictors, from eight-inch lizards to eleven-foot pythons. At first I believed, as most people do, that all of these cold-blooded creatures are equally cold hearted, but to my amazement some of the most moving of all of my experiences with animals have turned out to be with reptiles.

Snakes, I found out, can suffer from loneliness and depression right along with the rest of us, and be equally filled with joy and gratitude. Under the armored skin of the iguana beats a heart as fully capable of affection as that of any furry pet.

Many people instinctually shrink from reptiles as though there is something monstrous about them, perhaps in part because their cold-blooded slitheriness and armored scales are so genetically alien to us. Looking at them, we mammals find nothing warmly familiar with which we can identify.

Actually, seen from the reptilian viewpoint, we are the aliens—they have inhabited this earth for eons longer than we ourselves have. For more than one hundred fifty million years they dominated the terrestrial environment and it is estimated that six thousand species are now spread across all the continents of the globe.

Zoologists have classified reptiles into four main groups: snakes and lizards, turtles, crocodiles, and tuatara (reptiles that live mainly off the coast of New Zealand and look like lizards but are actually more closely related to dinosaurs).

The thought of the classification method makes a lot of people yawn, conjuring up endless lists of species with forgettable Latin names. But actually, aside from being our way of ordering life to make it comprehensible, *how* we classify animals is a wonderful mirror of western civilization's attitudes toward the natural world.

Only a few centuries ago in Europe, for instance, humans were considered to be at the center of the natural world and naturalists defined animals according to their relationship to man: edible or inedible, wild or tame, useful or useless. A little later, when a category was added that defined creatures as physically beautiful or ugly, the alien-seeming reptiles were instantly classed as foul, repulsive, and even evil.

We have a more enlightened attitude now and measure each animal's place in the world according to its own attributes and structural qualities. Television specials and nature books challenge our old prejudices: we learn how a snake shedding its skin is demonstrating a miraculous gift for self-renewal, that its restlessly darting tongue is not seeking to release venom but is testing the air for environmental signals, that the snake's seemingly pitiless gaze is unblinking not to terrorize its prey but merely because it has no eyelids.

Deeply ingrained reactions do not die quickly, however, and for many people just the thought of a snake is still enough to produce a little instinctual shiver.

Mythologist Joseph Campbell has an interesting explanation:

The serpent sheds its skin to be born again, just as the moon sheds its shadow to be born again. They are equivalent symbols. Sometimes the serpent is represented as a circle eating its own tail. That's the image of life. Life sheds one generation after another, to be born again. The serpent represents immortal energy and consciousness engaged in the field of time, constantly throwing off death and being born again. There is something tremendously terrifying about life when you look at it that way. And so the serpent carries in itself the sense of both the fascination and the terror of life.

It would appear that across the ages we've placed a heavy load on the slim back of the snake, a confusion of instinctual memories and cultural perceptions. As in the case of cats, we have reacted to the snake's special attributes by making him our emblem of both the fearsome and the divine, seeing him at once as the master of slippery treachery and the sacred incarnation of sexual, intellectual, and mystical power.

When I was a young girl back in that Canadian school yard, I thought I was afraid of a harmless little garter snake, but what was scaring me wasn't the snake at all. As soon as I experienced the actual snake itself, held its body in my hand, all my preconceptions melted away, leaving behind the reality of a very scared and very beautiful little individual.

Now, when I demonstrate my work with reptiles, one of the most moving things that happens is witnessing the way many people who fear snakes experience a similar change of heart. Seeing with their own eyes that snakes have feelings and are capable of relating with great delicacy and beauty to humans affects people strongly. Sometimes, I'm told, it even changes their outlook on nature, opening them to a new world of possibilities.

▷ Joyce

One of the most affecting moments in my life with animals was the recognition expressed to me by a Burmese python named Joyce (see photo insert). Joyce is eleven feet long and about the circumference of an average man's upper arm. She's very beautiful, with her light and dark brown body designed like a camouflage jacket and her diamond-like head. I met Joyce at the San Diego Wild Animal Park where I was demonstrating the TTouch at the twentieth annual San Diego Zoo-keepers Conference. At the invitation of Art Goodrich, historian of the American Association of Zookeepers and longtime keeper at the San Diego Zoo, I was to work on a number of animals, among them a reptile, a bird, and an exotic cat. The cat turned out to be Speedy, the hyperactive serval we met in Chapter Three. The reptile was Joyce.

I had already had the experience of working with an injured python at the Los Angeles Zoo and with many other snakes in various zoos around the world, and I was looking forward to learning more. I was told the snake was owned by people who had kept her as a pet until she grew too big to keep at home. Now she was on loan, one of the animals starring daily in the Park's educational Critter Encounters show.

Joyce was just fine for the show, I was told, but apparently, every year at some point she suddenly grew sluggish and miserable. Also, she suffered from a recurrent respiratory ailment.

The Critter show was headquartered in the midst of the park's lushly landscaped eighteen-hundred-acre spread. I waited to meet Joyce in the training area, a cement apron about twelve by sixty feet. A show was in progress down the hill from the training compound and I could hear the magnified voice of the lecturer-trainer echoing over the loudspeakers.

When Joyce was first brought out to me, she was placed on top of one of those round stools on which elephants balance their feet. She raised her head inquisitively.

Since the python was having respiratory problems at that time, my intention was to demonstrate how to work on her to prevent recurrence and to help relieve the congestion.

Gently, using a Raccoon TTouch (see page 244) no heavier than a three, I placed my hands on the smooth scales of her body in the area about three or four feet behind her head. Even this light contact, however, made Joyce very twitchy. And when I decreased the pressure to a one she still continued to twitch her skin. Like people with conges-tion and inflammation of the lungs, she may have had a sore back.

Switching to the flat-handed and tranquilizing comfort of the Lying Leopard TTouch (page 242), I slowed my breathing and kept up very light circles. As I concentrated on the snake, all verbal mental chatter vanished and the world became centered in my listening hands and the connection between us. Intuitively, my fingers moved to a position under her.

I picked her up and released her very gently at one-inch intervals using the rhythmic six-second hold-and-release pattern of the Belly Lift (page 258), a TTouch we usually employ on any animal in shock or with a digestive problem.

After a few minutes of work Joyce began to slide down from her perch on the stool. She moved in slow undulations, like a rope uncoiling, until she lay in the center of the encircling group of keepers, her body forming a shape like two joined Ss. Then, as we watched, she stretched out completely, straight as a stick, to her full eleven feet.

The keepers were amazed. Joyce lived in a small enclosure and was taken out once a day. I began to get an inkling of what might be causing the respiratory problem. A snake's lungs (or lung—pythons have two but some other species only have one) are not limited just to the upper part of the body but run half its entire length. When a snake doesn't have a chance to exercise or can't stretch out fully, the effect is somewhat like what happens to a human who is encased in a corset all day.

Joyce's lungs were probably in need of a good workout. So several assistants and I stood side by side and formed a line the length of Joyce's body (see photo insert). Like a human wave, one after another, we gently and slowly picked up and released her body in the Belly Lift and then repeated the action, though this time we very slowly moved the skin up just a quarter of an inch and then released it in what we now call the Python Lift (see page 254), in honor of Joyce.

After that we stood back to see what Joyce would do next. She raised her head, looked around at us, and then went slowly off for a luxurious slither.

As I watched her gliding across the unyielding cement, I suddenly recalled a video I had seen of a python swimming in the Amazon River, and soon we were all giving Joyce a bath. She loved it, and reached her head up as we splashed refreshing water all over her body.

Afterward, we formed our snake-treatment conga line again, each of us working on a different part of her with Lying Leopard circles. This awakened her torpid respiratory system and allowed her to sense her entire body, all eleven feet from head to tail.

So far we had worked with Joyce for forty minutes. I decided it was time to try the TTouch again on what had been the sensitive and especially twitchy area a few feet behind her head. This time she was perfectly quiet and didn't twitch at all.

As I moved over her in the Raccoon position, she curved her body and brought her nose around to my hands to observe the entire proceedings (see photo insert), hovering there with such obvious and intelligent curiosity that all we humans were fascinated.

Afterward, I sat back on my heels in front of Joyce and continued lightly working her head while we discussed her case. As I talked she began to slide gradually up my body until she reached my right shoulder, where she paused for a moment before moving very slowly and gracefully across to my left shoulder. From there she rose in a fluid motion until she came to rest with the side of her head against me just above my left ear.

I will never forget the sensation of her careful gentleness, the soft way she moved, barely stirring my hair before she pressed her head against mine with a pressure as light as a soap bubble. She hovered there, head to head, for what seemed like a full minute, sending me a clear and unmistakable feeling of thanks and acknowledgment. Nothing like that had ever happened to me. I was absolutely astonished, but there was more to come.

Two days later I was out on the stage, demonstrating the TTouch to the approximately one hundred members of the conference who sat facing me in chairs arranged beside a small artificial lake. Joyce was brought out. As soon as she was released from her carrying case she headed straight to where I waited in a kneeling position. When she arrived in front of me, she rose up before me like an Indian fakir's snake rising from a basket. She hovered for a moment at eye level, and then flicked her tongue against my forehead in a salute so light it was like the graze of a snowflake. It felt exactly like a gesture of greeting. Having done that, she slid back down and lay across my lap to wait quietly for me to begin my demonstration.

LEAPIN' LIZARDS, THEY'RE LOVELY

Animals vary, like people, and though I'd met numerous snakes before, Joyce was very special. The first reptiles with whom I was on familiar terms belonged to my second husband, Burch. When we were newly married he had two small boas, about four feet long, that he kept in large, glass terraria.

I felt sorry for them in their glass prisons and gave them the run of the house, something I wasn't exactly in a hurry to tell the average visitor. Caesar and Krono were quite elusive and I didn't really get to know them very well, but Burch also had a pet lizard who was about a foot and a half long and I used to let him out, too.

At that time, like most people, I wasn't aware that reptiles are capable of emotions, so though I liked the lizard, you could say our relationship was neutral. One night, however, as I sat reading, windows open to a summer twilight, the lizard suddenly emerged from behind the couch. He stared around sleepily for a moment, like someone waking from a nap, then lifted his head as if to sniff the wind, a connoisseur of atmospheric changes.

For some reason I continued watching him, not absently, but with my full and undivided attention. As I looked, my preconceived picture of "lizard" melted away. It was as though I was seeing him for the first time, a small dragon looking as mythic as any in fairy tale or legend, squatting there on my living room floor.

As I sat looking at him, he turned his head toward me as though drawn by the warmth of my thoughts. He returned my gaze for a moment, and then climbed carefully up onto the couch and positioned himself across my lap. At first he kept an interested golden eye cocked at me, but soon I could sense him relax and drift into a cozy drowse, companionable as a small snoozing dog. Wait a moment, I thought, I know about lap dogs, but a lap lizard?

Since then I've worked with quite a few lizards and have learned what truly marvelous gifts they have. Their scaly skin allows their bodies to retain water for long periods; some have a third eye on top of their head that measures just how long they may be safely exposed to the sun; many have the enviable ability of simply growing a new tail or leg to replace a mangled one.

As if these talents were not enough, the different members of the lizard family are endowed with more unique means of self-defense than James Bond: they can change color, squirt blood from their eyes, seal themselves into their lairs with sand, and inflate themselves with air like a balloon.

The latter ability belongs to the chuckwalla, a rare and small (six to eight inches) fellow who lives in the barren and rocky places of the American west and southwest. If disturbed while taking a nap on a sunny boulder, he scuttles into a fissure in the rock and puffs himself up until he's so firmly wedged that no predatory paw or inquisitive beak can possibly pry him out.

I worked on a chuckwalla once at the Syracuse Zoo. She had a spinal problem that had paralyzed her hind legs and therefore couldn't urinate or defecate for several days. The zoo personnel were worried that she was about to die of toxicity. "There isn't too much you can do to relieve a small lizard," someone said. "After all, there's no way to put a tube down its nose."

When I first saw her, the little blue chuckwalla seemed almost catatonic, eyes half closed, head low, belly hard and distended. I set her on a table, slipped my fingertips under her abdomen, and did tiny Raccoon TTouches and infinitesimal Belly Lifts with a number one pressure. After ten minutes or so—success. There she was standing in the middle of a puddle, relieved of a four-inch stool. We all broke into cheers, delighted, but not as delighted, I'm sure, as the chuckwalla.

The biggest lizard I ever worked with was a fifteen-year-old iguana named Mr. Big Guy (see photo insert), whose advanced age was beginning to make him slightly arthritic. The keeper wanted a demonstration of how to reduce the stiffness and prevent acceleration of the condition. Mr. Big Guy had a gray-green, three-foot body with scales that swirled in intricate patterns like exquisite beadwork. I had expected that I would have to restrain him at first—iguanas in the wild have a reputation for rapid departures, zooming along the ground with the front of their bodies raised like speedboats in high gear—but he was calm and trustful.

With the tip of one finger I made tiny circles all over his tail, up his spine, into his webbed toes, all around his head and mouth. Touching him amazed me. He was so receptive. Somehow you just don't expect delicacy and sensitivity from a reptile, maybe because that tough hide fools you, but as soon as I laid a hand on Mr. Big Guy, he closed his hooded eyes and lifted up his head, ready for a cell-to-cell talk. According to the latest report, Mr. Big Guy is now the longest-lived iguana in the zoo's history.

RATTLERS: WHERE THE WILD THINGS ARE

J. Allen Boone, in his eye-opening book *Kinship with All Life*, tells of a fascinating observation he made about rattlesnakes in the wild: "Almost everywhere I went," he says, "there was vicious and relentless warfare going on between white men and rattlesnakes . . . but I could find no such warfare going on between the Indians and the rattlesnakes. There seemed to be a kind of gentlemen's agreement between them. In all my journeys I never once saw a rattlesnake coil, either by way of defense or attack, when an Indian walked into its close vicinity."

How this could be did not become clear to me until 1966. Until that year I, too, was engaged in "vicious warfare" against rattlers. Like most ranchers, I had been taught from early childhood that rattlesnakes were killers. And what do you do with a killer? Why, you kill him first, of course, on sight, wherever and whenever you find him.

But in the summer of 1966 something happened that changed my prejudice forever. Went and I were living in Badger, California. We had moved our Pacific Coast Equestrian Research Farm from Moro Bay to a thirty-six-acre ranch in the foothills of the Sierra Nevada mountains bordering Sequoia National Park.

In the winter the hills were bright green and in the summer the wind riffled the dry oat grass on the slopes into endless golden waves. We took our students riding on trails that led deep into virgin forest where sunlight filtered down in cathedral-like shafts between the huge bodies of ancient redwood trees.

We had two barns, fifty horses, twenty residential students, and facilities for events for one hundred people. Our students were in the saddle four hours out of every day, seeking to master a variety of riding disciplines, as well as learning how to train horses of many different ages and breeds.

I used to go to the little town of Badger to do the shopping and errands, driving there on steep, winding mountain roads in my big Chevy van. One day when I was almost halfway there, I found myself staring through the windshield at a huge rattlesnake about six feet long who was trying to cut across the road in front of me.

Without a moment's hesitation, I slammed on the brakes, jumped out, and began heaving rocks at him from a distance of about four feet. Though I managed to hit him repeatedly, he didn't stop or coil to strike, but just continued traveling as fast as he could, heading for cover on the opposite side of the road. I kept on hurling rocks and hitting him until all of a sudden he simply stopped dead in his tracks.

I stared amazed as strange words entered my mind, like a radio frequency turning on in my head. "Oh, no," I heard, "she's really going to kill me." A split second later the snake had veered around and was coming straight for me.

I turned and ran quickly for the car, not because I was afraid (at four feet away I was in no danger), but because I was profoundly shocked. The inexplicable eeriness of the experience was totally disorienting. Alien words had entered my head and now, as I sat there in the car, I could suddenly see myself through the eyes of the snake. The mindless brutality of my actions disgusted and disturbed me. I grew more and more upset.

That was my first experience with hearing an animal's "voice." It was bewildering and filled me with self-doubt. Was I crazy? If I told the story would other people think I was crazy? What had actually happened? I kept the incident to myself. Later, I found out that I was not alone, that many people have such experiences and that almost all of them have the same initial reactions.

One of these people is J. Allen Boone. At that time, I had not yet heard of him, but a few weeks after my meeting with the rattler, I picked up his book *Kinship with All Life*. It gave me some much-needed corroboration and perspective, described other similar encounters he had experienced with animals, and detailed the remarkable relationship between the Native American and the rattlesnake:

> Could you look deep into the thinking and motives of the
> Indian, you would discover the simple secret of it all, for
> you would find that he was moving as best he knew how

in conscious rhythm with what he reverently called The Big Holy, the great primary Principle of all life, which creates and animates all things and speaks wisdom through each of them all the time. Because of this universally operating Law, the Indian was in silent and friendly communion with the big rattler not as "a snake" that had to be feared and destroyed, but as a much-admired and much-loved "younger brother" who was entitled to as much life, liberty, happiness, respect and consideration as he hoped to enjoy himself. His "younger brother" reacted accordingly. . . .

White people on the other hand have been taught to approach the snake with a degree of venomous loathing, horror and alarm that can only be answered in kind.

This wisdom made a deep sense to me and I decided never again to harm a snake. I asked that no one on the ranch kill snakes either, and requested that I be called if anyone did encounter a rattler in places like barns or pathways where it might be dangerous to people and animals.

I had not lost sight of the fact that a rattlesnake bite is serious, but when I checked at the library I was amazed to find that one's chances of being killed by a snake are considerably lower than those of being struck by lightning and that with proper treatment 98 percent of all victims recover. Several of our horses and dogs on the ranch had been bitten, but only after frightening or threatening a snake. Every one of them survived with no trouble other than a few days of swelling and fever.

Most bites, I learned, result from careless handling or failure to follow proper precautions in the field. (You should wear high leather shoes for hiking or climbing, watch where you put your hands when you climb, and use tools to overturn rocks or logs.)

It wasn't long before my new philosophy was put to the test. Our land, a blooming world of streams and thickets, of blond wheat-grass meadows and grandfather oak trees, was a haven for every creature that flew, walked, crawled, or slithered. So it was no surprise to me when, several weeks after my roadside encounter, I met another rattler on our cross-country riding course. He was large, a good five feet long with about eight rattles on his tail.

I went back to my car and took out two slats of wood three feet long and a few inches wide that I happened to have in the trunk. Snakes can reach to strike only one third of their own body length when coiled, so keeping a mindful distance, I placed one of the slats in front of the snake's nose and the other gently on the top of its head. Mentally, I told him I was sorry to disturb him but that I had to drive him down the field to take him to a spot where he would be safe.

The snake halted and looked at me without coiling or exhibiting the slightest sign of alarm and I could sense him evaluating the situation. Then he allowed me to gently herd him a good two hundred and fifty feet down a hill to a field where there was an empty fifty-gallon oil drum.

Again I mentally asked the snake to be patient with me. To my total amazement he didn't try to slither away or coil to strike, but waited quietly in place while I turned the drum upside down and put it over him. Once he was secured, I drove back to the main ranch to get a fifteen-gallon bucket with sides about two feet tall.

On the way back to the field I kept thinking, How am I ever going to get the snake into this bucket? It turned out to be surprisingly easy; I just laid the bucket on its side in front of the snake and asked the rattler to get into it, and in he went without the slightest hesitation.

When I picked up the bucket using one of the slats (I was still being careful of distance), he did no more than rear up and gently investigate all around the top before curling up in the bottom.

I drove a few miles down the road with the snake in the back of the van. As crazy as it sounds, he never moved a muscle the entire time. At a pleasant spot near a thicket I got out, turned the bucket over in the grass, and stood watching the snake glide off. When he got about twelve feet away, he turned, coiled, looked at me, and repeatedly made a sound I've never heard from a snake before or since. It was breathy, a kind of "thaaaaaah" sound.

For a while we just looked at each other while he made this sound and then I sat down. Gradually, I paced my breath flow to his and then it seemed as though there was nothing but the two of us sitting together in a moment of timelessness.

The bubble burst when I suddenly realized that twenty students were back at the ranch hungrily waiting for lunch. I stood up slowly, said good-bye and thank you to the snake, then watched him disappear into the undergrowth. I looked at my watch—thirty minutes had elapsed.

After this I began to be able to show other people what I'd learned. I found that the majority of snakes in the wild will ignore you if you stay at least twelve feet away, but if you are afraid or nervous and come within that perimeter, the snake will warn you back. All animals (humans, too) have a field of personal space that they do not like violated, but with snakes, as with other animals, there is a way of entering that space respectfully and harmoniously so that you are not perceived as a trespasser.

I had an assistant at Badger, Val Pruit, a person with great spirit and humor who would try anything once but who did hope that whatever it was wouldn't have anything to do with snakes. As Val's luck would have it, one day as she and I were up in the hills checking the jumping course, we came upon a rattler at the foot of an oak tree. It was summer and we had a lot of horses, animals, and young students around, so it was important to get the snake out of the area.

With Val hanging back behind me, I walked slowly up to within eight feet of the snake and stood quietly to make contact. I motioned Val to come closer because I wanted to ask her to go and get a bucket, but as soon as she crossed the snake's safety perimeter he reared up and rattled furiously. He had instantly picked up and mirrored her fear. When she hastily withdrew he calmed right down again.

The best way to make contact with a snake is not with a lot of busy words going on in your head. Instead, what you want to do is to project an attitude of respect and admiration.

I described this to Val and asked her to slow her breathing, soften her eyes, and look down at the ground just as she would when approaching a wild horse that won't allow itself to be caught. (Unfocusing the eyes and avoiding eye contact is a prime rule in dealing with frightened or aggressive creatures who feel threatened.)

Gradually, as Val concentrated on these points, she was able to come in closer and closer without alarming either the snake or herself. At last she stood beside me, only eight feet from where the rattler lay, uncoiled and calmly interested in us. We stood silently enjoying the moment—the snake, the leaf shadows that played over our faces, the wind that sighed through the tall grass and moved it in ripples around us.

WHAT DO PET REPTILES NEED?

Reptiles in captivity respond in much the same way as those in the wild—they need understanding. Actually, they need even more than that, because they are in an unnatural situation for them. If your reptile is aggressive, there's a good chance that poor environment and lack of contact have something to do with it.

Shortly after I first began developing the concepts of TTEAM, Eleanor Elter, the niece of my teacher Moshe Feldenkrais, signed up to attend a six-week horse clinic with me. She felt that working with horses would teach her a lot about human beings and she also wanted to work with her snake, a rosy boa whose behavior was troubling.

Eleanor, who was studying herpetology at California Polytechnic University, was worried about Rosy. The snake was depressed, had lost her appetite, and would attack whenever anyone tried to reach into her enclosure tank. Although I only worked professionally with horses at that time, I thought Rosy would slip easily into the program and I was delighted to take her on as my first non-equine case.

I'd be depressed, too, if I were a snake, I thought as I looked at Rosy's drab glass tank. There was nothing in it except a dried twig and some sand. So the first thing we did was to give her an interesting branch to climb on and a large rock for her to creep under for privacy and the chance to shift back and forth between light and shadow whenever she chose. We put a bit of old carpet on the bottom so that she would be physically stimulated by two different textures.

Eleanor brought her to the training sessions and various members of the group worked with her, handling her often and making her a part of the social life of the group. After three weeks she regained her appetite and was active and happy in her enclosure. But the final test came at the party with which we marked the end of the six-week course.

Rosy was invited, too, and she loved the celebration. She joined in the dancing, draped around one person's neck, wrapped around someone else's waist, passing from one to another of us like a link in a chain.

Thinking about Rosy and Joyce and Mr. Big Guy, about rattlers and lizards and all the extraordinary qualities of the reptile family, I remembered something I once read by naturalist Henry Beston:

> The animal shall not be measured by man. In a world older and more complete than ours, they move finished and complete, gifted with extensions of the senses we have lost or never attained. . . . They are not underlings; they are other nations, caught with ourselves in the net of life and time, the splendor and travail of the earth.

8

Down on the Farm:
The Cycle of Life

The evening milking was a restful moment in the day. Men and women, tired after their work, slapped their little milking stools on the ground beside the cows. Having made sure that your block [stool] was secure, you sat down with a bucket between your knees and rested your head against the soft, silken flank of the cow. Then, wrapping your fingers around the cow's warm teats, you milked to a steady, soothing rhythm. At first the milk hit the bucket with a sharp metallic sound but as it filled it mellowed to a drowsy hum and the cold bucket grew warm between your legs.

—Alice Taylor
To School Through the Fields:
An Irish Country Childhood

Growing up on a farm is very instructive in a number of ways. For one thing, I never needed to attend any classes in sex education. By the time I was six years old, I had already witnessed a rich weave of calving and foaling, of breeding and mating, of death and survival. The triple mystery of sex, birth, and death was as natural a part of the round of life as dawn and dusk. So, too, was the sense of protectiveness and responsibility that came with "owning" animals.

For my ninth birthday my father gave me a pedigreed holstein calf.

I loved to tell people his registry name—Highvale Architect Fairchild Dale—it rolled so sonorously off my tongue. Highvale was the name of the sire, Architect was the family line, Fairchild was the mother, and Dale was the calf's personal name. In moments of annoyance or love, I would call him by all four names at once.

After a few months the young bull grew to exceptional size with fine strong legs, a straight back, and a clean top line. My family and the neighboring dairy farmers encouraged me to enter him in competition at the agricultural fair in Edmonton. I thought about it and decided I liked the idea. I wanted Dale to have his own ribbon to add to the impressive reputation of the family of Architect. But while I knew about riding in horse shows, I knew nothing about presenting cattle at fairs. I imagined there would be no more to it than parading my calf past the judges.

Arriving in the dairy section of the fair was a shock. All the cattle looked as clean as new stuffed toy animals, as if they were all made of satin with gleaming hoofs that had never seen mud, and tails that were trimmed and neat as combed tassels. The aisles were crowded with people talking and laughing, and everyone seemed to know each other. I walked up and down the rows of straw-filled pens getting more and more discouraged and wondering how I could ever transform my calf into anything that resembled these bovine movie stars.

At last a man whose two butterscotch guernsey heifers were enthroned in the space next to mine took pity on me. He had closely cropped gray hair and wore a new white shirt and jeans especially pressed for the occasion.

"First time at show?" he asked. He leaned on the wooden stall partition and looked Dale over. "Nice calf," he said. "Need some help with him?" I felt like throwing my arms around this savior's neck but instead I merely nodded vigorously. He came around into my pen with brushes, hoof polish, and clippers, and then for the next hour he showed me how to groom a calf for show.

Dale took second place, and as I led my calf around the ring in the winner's parade, almost suffocating with the pride of the moment, I kept looking for the man's face among the spectators', but he wasn't there. I'm sure he never thought twice about what he had done for a scared little girl.

BIRTH

Farm animals probably more than anything else in my life have taught me the circular nature of birth and death. In 1959, my husband Went and I were partners in a thoroughbred breeding farm at Hemmet, Southern California. Besides ninety brood mares, four stallions, a band of regular mares, a family of Great Danes, and several Siamese cats, I was also responsible for thirty ewes and a ram. It was the beginning of my shadow career as a shepherdess—for some inexplicable reason I have often found myself minding small herds of these woolly characters.

My first lambing season at Hemmet was disturbing. Sheep are panicky, passive creatures and they tend to give up easily: I've seen ewes fall into shallow ditches and just lie there helplessly pawing the air, absolutely convinced they are trapped. At that time I had never experienced lambing time. Most of the ewes were giving birth to twins and triplets, popping them out on the ground. The lambs would lie there for a moment, stunned by their arrival, and then rise up on wobbly legs to nurse. If they didn't get the hang of the nursing after ten minutes or so, they would lie back down again and simply die.

That's just how they are, I was told by the sheep experts I knew, some live and some don't. Being stubborn, I had other ideas. I walked around the pasture and every time I found a lamb that was wilting away on the ground and wouldn't suck, I picked it up, shook it a little and held it by the throat, forcing it to struggle for breath. In this way I actually choked quite a few of them back into life. I can imagine the scene from an outsider's viewpoint—a young woman striding around a field choking newborn lambs? Very strange. But it was better than doing nothing and it did work some of the time.

Had I known how to use the TTouch then, I'm quite sure I could have saved most of the newborn flock. The method is simple: you make small Raccoon circles on the lips and inside the mouth on the tongue and the roof of the mouth. Then you thoroughly work the ears, sliding your fingers from the base to the tip. These actions activate the stalled sucking reflex almost instantly and give the lamb a burst of energy, a magic jump start back into life.

Hemmet was primarily a thoroughbred breeding farm. Brood mares foal January to April, and during those months I found myself very busy in the role of midwife. For me, assisting labor was always a very special time of elation mixed with fatigue. With so many mares giving birth, I was up at all hours, and sometimes got little sleep for days on end.

Often when I had several mares all due at the same time, I would have to bed down in the barn, but I loved an excuse to spend the night lying in the sweet-smelling hay, listening in the dark to the comforting sounds of the night, the chickens rustling in the straw, the *chirr* of crickets, the sighs of the horses. Some mares would snore, others kicked and whinnied in their dreams. In their sleep as in their waking hours each one was different, a distinct personality.

To me, there's something almost primordial about the feeling of a nighttime birth, the softly lit warmth of the stall like a womb itself in the encircling darkness, the rustles and munchings of life going on around us, the trust and appreciation of the mare in her labor, a moment of intimacy that is like no other—and then the gift, the foal.

Things do not always have happy endings, of course, and the possibility of death or injury is present at every birth, like the outline of the moon's dark side dimly perceivable against the bright.

I remember one filly who was born with an incomplete closure of the navel. As I waited for the vet, I held her in my lap, desperately trying to hold back the protruding intestines with my hands. She was a fierce little foal and I could feel her strong will for life, her desire to fight for it with each struggling breath—she would have been an exceptional filly. To make matters even worse, the mother was standing beside us, looking on anxiously, but the foal didn't make it.

I think that the births and deaths of animals close to us allow us to experience thoughts and feelings that are normally kept deeply buried, and the foal's death, the sight of her mother nuzzling her lifeless body, tore at my heart. At the same time, as someone who had grown up on a farm, death to me was something you just accept, a given like the seasons. I knew that acorns fall from the trees to the ground and not all of them are destined to become oaks.

TALKING WITH A FETUS?

Thirty years have passed since those California days, and I live in New Mexico now. Not many miles away from me, in the one-store desert town of Galisteo, my good friend Priscilla Hoback owns Quartermill Farm.

"We run the gamut with our horses," Priscilla says. "We live with them, we ride, show, board, and teach people on them."

Though she is obviously a serious and passionate horsewoman, Priscilla's world actually embraces a field of knowledge as wide and exotic as the desert that rings her adobe house. She is a world-renowned artist and potter. The walls of her many cool rooms are crowded with portraits of friends, with minutely observed drawings from nature, and with landscapes of near hallucinatory power. Song birds flit in a huge aviary, bunches of dried herbs hang overhead, carved totemic animals and ancient pottery reflect her feeling for her Native American heritage.

I often take the drive out to Quartermill Farm through the desert to visit with Priscilla. Her fine-featured face beneath a colorful head wrap welcomes me, dark eyes a little guarded but warm. She is usually excited about one of her horses and takes me straight to the barn.

We first met through the TTouch. Priscilla had been reading about it in a horse magazine and had tried it out. Eager to learn more she attended one of my clinics. Since then she's done more than practice the TTouch therapeutically on her horses—she's used it in creative research of her own, coming up with work that has fascinated me because of its wider implications.

Priscilla was especially interested in using the TTouch on animals, particularly foals, in utero. She noticed that certain seemingly acquired characteristics, like refusal to allow the ears to be touched, actually seemed to be inherited.

It was one of her mares, Roseata, who started her thinking about this. "Roseata was probably not treated very well as a yearling," Priscilla explains. "I joke that the reason I keep her is that nobody else would put up with her. She's a really emotional horse, bossy, particular, and very intelligent. She's opinionated and doesn't like anyone but me. If she were a woman she'd be a very high strung and difficult female to deal with, a real queen. And before we used the TTouch on her, she had supremely sensitive ears. She wouldn't let you near them without a fight to the finish.

"Then along comes her son Chakra, and he's got exactly the same ear problem, although he certainly had no history of abuse. The opposite—Chakra was treated like a little prince from the moment he was born. Never was a horse more wanted or loved, yet he was totally terrified to have anyone go near his ears. He was bad as a horse can get."

After Chakra was born, Priscilla and I discussed the case and I told her that I'd been speculating for years about the inherited character of certain tensions and difficulties. I believe that quite possibly certain traumatic experiences of the parents or the grandparents are passed down in the form of cellular memory.

By putting new information into the cells through the TTouch, by giving the cells the opportunity to release the programmed genetic tension and the expectation of fear that it carries, it seems to me you have a chance at changing the foal's behavior.

But wouldn't it be easier, I speculated to Priscilla, to practice visualization and the TTouch on the fetus in utero rather than on the fully formed baby after it is born? Wouldn't the fetus be more malleable, more open to work, on the cellular level?

Priscilla was intrigued and began to put this revolutionary hypothesis to the test.

"I worked on my pregnant mares, visualizing the positive result I wished for in the comparable body part of the fetus," she says. "With Roseata I worked the ears while visualizing the ears of the fetus; with another horse I worked a weak, short neck while seeing a strong, arched one forming in the unborn baby. For a calm temperament I worked the TTouch over a nervous mare's entire body.

"Everyone who breeds horses has a lot of hopes and dreams going anyway, a lot of mental images, so all I'm really doing here is creating a focus for those images by combining visualization with a corresponding physical action."

Priscilla has bred three foals using this method and has noticed that all of them displayed the same unusual characteristics. None of them needed the usual preliminary gentling and convincing to train them to halter; from the moment the halter was placed on them they just seemed to know what to do. They are exceptionally eager to participate in the training and to do anything that is asked of them and all three are outstandingly intelligent as well as beautiful. Since these experiments, I've recommended to a number of breeders that, as they work with the TTouch on their pregnant mares, they keep in mind an image of the foal as they would wish it to be. Although no conclusions can be drawn from our limited data as yet, the method is worth noting and it certainly can't hurt.

PRACTICAL TT: HELPING A DIFFICULT BIRTH, FROM CALVES TO KITTENS AND PUPPIES

In the physical process of birth, TT is most often applied as a practical aid to help labor, to ease pain, and in some cases, even to save lives.

Allison is a woman who knows this first hand. She and her husband Doug owned a beautiful, creamy charlet heifer who gave every indication of being a prize addition to their breeding stock. Doug had spent a great deal of time, money, and care on her and had high expectations of her first calf. The heifer, however, was having a very hard time of it, as is often the case with first births.

Hour after hour passed with the poor creature exhausting herself, straining and pushing—doing her best, but to no avail. It was terrible and frustrating to stand helplessly by, and Doug and Allison grew heartsick. The calf was exceptionally large, and Doug and Allison became increasingly frightened that neither the heifer nor her calf would make it. Finally, the panting creature threw back her head and gave up in despair, lying there as though already dead.

Doug, who had birthed many heifers, realized to his dismay that he was left with no choice but to bring out block and tackle, attach the calf in utero by the legs, and winch it out of its mother's body. This technique, used only as a last resort, sometimes works but usually leaves the mother paralyzed or severely traumatized. More than likely her breeding days will be over, and the calf subjected to this treatment almost always dies.

Both Doug and Allison were distraught at having to use this last resort. As Doug was about to go off and get the equipment, Allison put her hand on his arm. She had recently been reading cases in the TTEAM newsletter and had seen a TTEAM video on how to use TT with mares who were undergoing difficult labor.

"Why don't you just give me twenty minutes more to try the TTouch," she said to her husband. "Maybe we can get her to start her labor again."

Allison began working on the heifer's ears as the first step in reactivating the animal's flagging energy. Around the base of the ear is the Triple Heater acupuncture meridian, which affects the respiratory, digestive, and reproductive systems. This area is especially effective in stimulating labor, and Allison worked the area thoroughly with small circles and then slid her fingers up from the base of the ears to the tip, where she made sure to pay particular attention to the shock point (pages 54, 121, and 265).

After working the ears for five minutes, Allison moved to the cow's pelvis, where she used flat Lying Leopard (page 242) movements. Then, reaching both arms beneath the cow's lower body, she performed slow lift-and-release motions like a big Belly Lift (see page 258). This shifting around of the weight of the calf will not only recharge the mother but will also help to get the calf moving again, too, because when the mother stops trying, the calf tends to give up as well.

After fifteen minutes of concentrated work Doug and Allison were overjoyed to see the heifer stirring. Clearly, she had made up her mind to go back to work. She went into labor once more and finally produced a very large and splendid baby, the Arnold Schwarzennegger of calfdom.

These methods are applicable not only to cows and most other farm animals, but to your household companions as well.

▷ *Kittens and Puppies*

If your family cat has never had kittens before and seems nervous or afraid, it will help her a lot to feel your presence and participation. Gently work her ears and around her pelvis, and in case you've never witnessed this miracle before and you feel nervous, try a little work on your own ears, too.

Bitches who have never whelped before and who are having difficulty with labor and birthing also benefit from ear work. Try a number four pressure Lying Leopard TTouch around the pelvis and gentle Belly and Abdominal Lifts.

Show dogs are often particularly nervous and stressed at whelping, sometimes because they were taken away from their own mothers too early, sometimes because the nature of their work makes them tense. In cases where the bitch is particularly anxious and unhappy during whelping I find adding toning and breathing to the TTouch can do wonders.

Most dogs are already very aware of your aid in being there and will easily fall into your breathing pattern, especially if you slow way down and breathe audibly. Slow, rhythmic breathing alternated with soft toning will encourage and calm your dog and take the fear out of what is happening to her. (It helps you, too, because unless you've been present for many births you might be a little nervous yourself.)

Some first-time mothers have no idea what is going on. I remember a mare I had who turned and looked around behind her to see the baby and was so shocked she leaped up and ran away. I had to catch her to introduce her to her foal.

In any case, whatever the animal, I've always thought of the event as being one of the best celebrations you can possibly attend.

▷ Cow Strategies

Bovine means cowlike or calm and placid. Well, yes, much of the time cows do live up to that reputation, but remember the old nursery rhyme about the cow jumping over the moon? Whoever wrote that probably had experience milking young heifers.

Young cows can be nervous at milking time. When this happens they will kick out, swish their tails, hold back their milk entirely, or have a difficult time letting it down. Most cows today are part of huge dairy herds and are milked by machine, a practice that engenders its own stresses, but there are still plenty of small farms in countries other than the United States that use the good old-fashioned three-legged stool-and-bucket approach.

An Austrian television producer asked me to give a demonstration of the TTouch on a program about veterinary problems. My animal clientele was to include a dog who bit, a cat who was tense, a horse who spooked, and a cow who kicked when being milked. With fifteen minutes for each animal I felt like an express-train conductor on a tight schedule.

The cow, Gretchen, was a caramel-colored jersey with such a sweet, innocent look in her big brown eyes that it was hard to imagine her doing anything more aggressive than munching on wildflowers. Milking, however, made her very edgy. As soon as the farmer laid hands on her she would start kicking and thrashing her tail so vigorously it would smack him in the face. Milking time was far from a peaceful bucolic ritual for this farmer.

The traditional solution for such rambunctious cows is to hobble them and tie down their tails, but Gretchen was too nervous even for this. She would continue to struggle, which is why she had been chosen to appear on the program with me.

In examining Gretchen I found that being touched generally made her very jumpy, so I got out my wand and stroked her on the chest until she quieted down. Within a few moments I was able to stroke her back but then, when I moved to her hindquarters, she began lowing her objections, moving away from the wand and kicking out.

If you touch an animal on both legs at once with two wands the normal signals to the brain are interrupted and the animal can lift neither leg. I stroked first the backs and then the fronts of both of Gretchen's hind legs at the same time. After she accepted this (with a rather bewildered expression on her face, I must say), I went on to use the back of my hand to circle her sides, flank, and pelvis. The back of the hand is much less threatening to a nervous animal than the palm. The connection is less intense and the animal can feel more trust knowing you are not in a position to grab her.

After she stood calmly for this I graduated to doing Lying Leopard circles on her back to distract her while I quietly moved down with my other hand to the bag and udders. In a few minutes I could stroke the bag easily with the back of my hand, pushing upward, which is less threatening to a cow, and then softly stroking the teats.

Gretchen stood quite still for most of this and as soon as she seemed to grow restive again I distracted her by working on her back with one hand and on the bag with the other. At the end of the session, Gretchen and I were serenely able to deliver a few squirts of milk into a bucket, while the studio cameras rolled and Gretchen became a momentary TV star.

Of course, I can't claim that Gretchen went home and became a model of bovine deportment after that, but it was a start. At least the farmer wouldn't have to send Gretchen to the butcher in despair, which is what happened to difficult cows like her in the old days.

The TTouch has also been successful in treating a mysterious phenomenon called "tingle voltage." Most people come to TTEAM because they're having problems with their animals, and as a "troubleshooter" I've become fairly used to dealing with the unusual. Tingle voltage, however, is one step beyond unusual to me.

In many areas an explicable and invisible stray electrical field occurs that somehow "charges" the cattle. It strikes at random and can be measured with a voltage meter held to the affected animal. Cows in shock with tingle voltage have a problem letting down their milk. To get them to do so farmers resort to chemical injections, which relieve the poor cows but also contaminate the milk and make it unsaleable.

A dairy farmer in Ontario wrote to our TTEAM newsletter to tell us he has successfully adapted the TTouch to relieve the symptoms of his "electric" cattle. At milking time he now makes small circles over their bags and teats and finds that within five minutes they are ready for milking.

What apparently happens here is that the stress caused by the charge is reduced, allowing the cow to relax enough to interrupt the cycle of charge and recharge.

Stray voltage is thought to cause depression, migraines, cancer, and other diseases in humans, so farmers and cattle ranchers in "electrified" areas might do well to work with the TTouch on themselves as well.

DUCKS, CHICKENS,
BARNYARD MAYHEM, AND MOZART

Chickens are amazing creatures. Many people think they're too dumb to understand anything but egg laying and crowing, but they can be very loving and communicative. I even know of a rooster who likes to cuddle up with people and who thoroughly enjoys classical music, especially Mozart.

His name is Chee-cawn and he lords it over the yard and stable of a lady named Sarah. Admittedly, nothing to do with Sarah is ordinary: her own exploits include riding wild horses bareback with the legendary cowboys of the Camargh region of France, driving a horse and carriage in New York's Central Park, and journeying on horseback from New York to New Orleans.

Still, Sarah insists that any rooster will be as friendly as Chee-cawn if treated with patient respect. She reports that Chee-cawn comes when called, poking his wattled, rainbow-feathered head around the corner of her barn to see what's up. While she sits on the threshold of the barn door he pecks about her feet with studied casualness before jumping with a flutter onto her lap. Once ensconced, he snuggles his head up into her neck right under the ear and makes small chortling noises from deep inside his throat, occasionally rearing back his head to give her a lovestruck look.

The rooster became a Mozart fan as a result of illness. One afternoon Sarah found him in the hayloft listlessly sitting on top of the feed barrel, eyes closed and tail feathers drooping. He was dehydrated and very hot to the touch. The vet, when he came, diagnosed poisoning of some kind, and for the next two days the rooster lay in the hay nest that Sarah had prepared for him. He had no interest in life, worms, breadcrumbs, or even hens.

Sarah liked to listen to her tape recorder while mucking out the stalls and washing the barn floor. She was playing Mozart's *Eine Kleine Nachtmusik* when she remembered that she hadn't checked on Chee-cawn for a while. Taking the tape with her, she climbed the stairs to the hayloft. At the sound of the music Chee-cawn opened his eyes, lifted his head, and feebly fluttered his wings. He seemed to enjoy the music so Sarah left the tape playing beside his nest.

The next day she went upstairs with another Mozart tape. Chee-cawn greeted the master's opus by getting to his feet and losing himself in a dance of total abandonment, ducking his head up and down and hopping ecstatically from foot to foot. Seeing that he was a music fan and that the stimulation was helping in his recovery, Sarah tried other tapes—Beethoven, Mahler, and Bach. He hopped with pleasure to them all, but continued, even after he had recovered his full health, to reserve his best dancing for Mozart.

Experiments have shown that plants react to music (they prefer classical) as well as to talk by growing healthier and taller. Clearly the benefits of music are deeper than we intellectually or scientifically understand. Music is extraordinarily helpful in conjunction with the TTouch in soothing any animal that has been traumatized or who is sick or nervous.

Like Sarah, many people find themselves developing a surprising special relationship with a particular farm animal. I've heard of pet chickens, pigs, calves, and ducks. I myself was bewitched by a motherless lamb named Junior who wound up sleeping in my bedroom. Just like the nursery rhyme about Mary and her lamb, Junior followed me everywhere.

A woman I know who runs a small "gentleman's" or should I say "gentlewoman's" farm found herself adopted by a duck. She quite enjoyed this state of affairs and so was horrified to come home one day to find that her pet duck had been badly mauled by a dog. The duck was probably not aware that this stray wanderer was not like the friendly and well-trained farm hounds to which she had become accustomed.

She was severely injured and it took four hours of working the TTouch on her head and body before she even opened her eyes, and a week of half-hour TTouch sessions three times a day for her to recover.

If you have a turkey, duck, or chicken that has been trapped in a corner and mauled by a dog, or a hen that has survived a raid on the henhouse, don't be discouraged if the bird looks limp and almost dead. Use the Abalone TTouch (page 248), moving both hands very quietly, slowly, and lightly over the feathers. Concentrate on giving warmth and contact. Then begin making the smallest of circles with your forefinger under the feathers and around the ear hole (for shock). Keep your TTouch as gentle as you can and don't be surprised if it takes a while for the bird to show signs of life. It can take as long as half an hour or more to see a response. (See the chapter "Feathered Brethren" for more information on handling birds.)

STRONG BACKS AND STRONG WILLS:
BURROS AND LLAMAS

Burros have a reputation for stubbornness. It's true—these cute little fellows with the fuzzy faces and the cuddly looks can demonstrate the will of a world-class dictator. They are masters in the refined art of passive resistance, and a lot of frustrated people wind up acting like cavemen when they try to wrestle their burros into submission.

Because burros are small, people think they can push and pull them around. Actually, the more disrespectfully burros are treated, the more they will harden their resistance, and the more they resist, the more they try the tempers of their owners.

I'm sometimes asked how I can stay patient even when an animal behaves in a frustrating way, but I don't see it as patience. Losing your temper with an animal is probably the same as feeling that you've run out of alternatives or "come to the end of your rope," and I find I can stay calm because I know that if one thing doesn't work I always have a number of other "ropes" I can try.

Resistant animals are great at teaching you how to stay cool, and burros are particularly well suited for the job of people training. Yet, even burros will peacefully cooperate once they've been shown what it is you want and how to do it for you.

One of the most frequent complaints about burros and donkeys is that they are very difficult to lead. It's as though they have some arcane ability to grow invisible roots that anchor them firmly to the spot. To get a donkey to realize that he doesn't have to be a tree, use the Dingo (page 271); first stroke his back with a wand or flexible stick, then bring him forward with your left hand on a lead chain around his nose while tapping his hindquarters and hocks with the wand in your right hand.

Another common problem is trimming the feet. Some burros, especially youngsters new to the farrier, will lash out and fight rather than allow him to touch a single foot. The TTEAM approach for this challenge works for horses as well as mules and donkeys. Begin by stroking the legs with the wand to calm the nervous reaction and to accustom the animal to the idea that a touch on the legs doesn't automatically spell danger. Then do Python Lifts (see page 254) and small circles up each leg. After fifteen minutes the animal will more than likely permit you to lift the leg with no problem at all.

All of these techniques have received a thorough and successful testing on subjects that are about as hard to train as you can find—wild horses and donkeys straight from the holding pens of the United States Bureau of Land Management. The BLM conducts periodic roundups in areas that have been overgrazed in the West and Northwest and administers a program allowing people to rescue the captured animals through free adoption.

I know of quite a few people who have adopted through the BLM program and have used TTEAM to calm and manage their animals, finding it a much faster and safer way to train them than any other.

Too bad there are no adopt-a-llama programs, although if there were, the supply would not be able to keep up with the demand. Llama ranching has become a rapidly growing industry in the United States as North Americans find out what South Americans have known for centuries—that llamas are wonderful creatures who will not only carry your burdens but will furnish you with exceptionally fine wool.

Another reason for the run on llamas is that people find them exotic and intriguing. Because they are large animals they don't seem like just another pet, yet owning a llama is much less complex than owning a horse. Too, llamas have a reputation for being highly intelligent, headstrong, and individualistic, a reputation they richly deserve.

My first encounter with a llama was actually potentially dangerous. It was in the days before we had developed TTEAM, when I was still living in California, operating the Pacific Coast Research Farm with my husband Went. My friend, the Hungarian countess Margaret Bessinger, had invited me to her spectacular spread in Montana, a ranch first developed by the famous "Anaconda Copper King" Marcus Daly.

Margaret impressed me tremendously, and not only because she was a highly skilled horsewoman. She was a generation older than I, and was equally at home at a black-tie dinner party or sitting around the kitchen table over a pot of stew. Margaret could make something special out of any activity, from fixing cheese soup to taking a bath. She spoke four or five languages, often mixing them together in a colorful scramble, and she had a great sense of style in which hats played an important role. On the day I met the llama she was wearing her favorite Stetson.

We had gone for a walk to the pasture because she wanted me to have a look at her herd of superb Hungarian horses, and also at a llama she had just received as a gift from a male admirer. A rather odd present, we agreed—the llama was a big male and very aggressive.

"Do you think my friend was trying to tell me something with this gift?" Margaret asked. We were standing at the pasture fence and laughing when the llama came right up to the rail to examine us. I don't know what he saw, but, quick as a flash, he reached over, whipped Margaret's hat off her head, and dropped it onto the ground on his side of the fence.

Without a moment's thought I climbed over the five-rail fence and quickly bent over to retrieve it. Not quickly enough, however. The llama charged me ignominiously from behind, ramming me with his chest as is the habit of attacking llamas, and knocking me to the ground.

The next thing I knew I had reached back, grabbed him by his snakelike neck, and brought him down to the ground beside me, holding him there by his head. I didn't know what to do next, so there we stayed in a strange impasse, neither of us moving a muscle. After what seemed like minutes, I let go and he jumped up and ran off. Looking down, I saw the Countess's hat lying there, flatter than an old tire.

Actually, I was just plain lucky, because I later found out that once a llama knocks you over, the next thing he usually does is lie down on top of you to pin you. A big six-hundred-pound llama like the one I tangled with would have crushed more than just my pride.

I've worked with many llamas since that first skirmish and have found that Margaret's hat stealer was not exhibiting normal llama behavior. He was suffering from what in llama circles is known as "the berserk syndrome," in which a llama will suddenly and inexplicably begin to attack humans and knock them down.

The first thing to do for a llama with this affliction is to lower the animal's head, as you would with any aggressive animal. Like the rattlesnake who rears up before striking, the horse who throws its head high when startled, the nervously barking dog, or the bristling cat with its head high and stiff, once the llama's head comes down the behavior changes dramatically (for instructions see pages 114–15). The animal will be ready to work with you and you will have achieved dominance without using force or demanding submission.

With a "berserk" llama we then follow up with the TTEAM exercise called the Journey of the Homing Pigeon (page 272), which, by placing the llama between two people, prevents him from attacking.

174

The first time I actually worked with one of these sensitive and beautiful creatures was when Samanda Berman, a neighbor and friend in New Mexico, brought her llama to a horse clinic. Samanda had a really unique problem: her llama was constantly spitting on her Arab gelding, flecking his beautiful white coat with greeny yellow spittle. The two animals were kept in the same pasture, but the llama was deathly afraid of the horse and spitting is a defense mechanism for a threatened llama. Most people think llamas spit out of aggression, but the basis of this truly ingenious and effective repellent is nearly always fear.

Samanda took me over to the horse field to witness the proceedings and sure enough, every time the horse approached within six feet or so, up would go the llama's aristocratic head and a second later he would let his defense arsenal fly.

We worked the llama and the horse together for one week, using a three-step process for half an hour a day. First we lowered the llama's head and worked his ears to calm him down, quiet his nervousness, and to enable us to work easily with him. Then we brought the horse to within eight feet of him, and with one of us working the horse and the other the llama, we did the TTouch over the entire body of each animal.

The next step was to work the two animals in hand together to give them a sense of kinship. We led them, side by side, over and through obstacles, and every time the llama reared his head, we lowered it again, toning to him and reassuring him by stopping and administering one of the soothing TTouches, the Abalone or the Lying Leopard.

By the following week, while you still couldn't call them bosom buddies, the llama was much more comfortable with the horse and had completely stopped his spitting. Needless to say, Samanda was delighted to see her horse's true colors emerge again.

At present, there are thirty thousand llamas being raised in the United States. The number is growing and llama owners are passionate aficionados. I wasn't really aware of the magnitude or dedication of the llama "industry" until I met Marty McGee, a weaver who uses the wool of her llama herd to hand loom blankets, sweaters, and coats. Her colors run from natural, tawny hues to the kind of pinks and purples that remind me of evening skies.

When Marty first phoned me we had never met, but she had heard that I was teaching a weekend TTEAM seminar at a college close to her home in Dunbar, New York. She called to ask if I would come up after the course and look at several nervous llamas in her herd. Normally, after completing a two-day clinic, the thought of driving an extra few hours out of my way held no appeal, but llamas with behavioral difficulties in the mountains of upstate New York? Sure, Marty, I heard myself say, I'd love to come up and see your llamas.

Marty's house is in a pastoral area of wooded hills, fields, and stone walls. The roads are lined with elms and oaks and rural delivery postal boxes. Marty herself turned out to be a tall woman with shiny dark hair and a pleasantly direct manner. Her living room was dominated by a fieldstone fireplace, a large loom, and piles of colorful skeins of llama wool.

Owning a llama, she told me enthusiastically, is a little bit like owning a unicorn. "When you watch them at play out in the fields," she said, "it's pure magic. They have a prancing gait so light and graceful it's somewhere between flying and dancing."

We went to her barn, where the "problem llamas" were waiting. The first two were just yearlings, fluffy white clouds with small heads and big, soft deerlike eyes. They were more difficult to catch and halter than llamas normally are, she said, and she wondered whether TTEAM might be helpful.

As every llama owner knows, llamas are quite aloof and they have a thing about letting anyone touch their heads—they absolutely can't stand it. Unless we could get these little fellows to allow the TTouch all over their heads, however, it would be very difficult to gentle them. What to do?

First I took a quarter-inch rope and tied it around the llama's neck, securing it with a bowline knot. The bowline is the only safe way to tie a knot around the neck of any animal you are training because it's the only knot that can be quickly and easily untied.

At the first touch of my hand on his head the poor little guy plunged away in total terror, so I began the TTouch on his long, graceful neck instead. Then I went back to his head, but this time, seeing that the touch of a hand was still too direct, I leaned my face against him and made soft circles on his cheek with my forehead. When I saw that he could tolerate this less electric contact, I used the back of my hand around his mouth until finally the llama was quite happy to permit me to do the full TTouch on his ears and even on his lips and gums. The whole gentling process took less than thirty minutes.

Llama owners and breeders attend wonderful events called llama jamborees, where the goings-on are a kind of cross between a rodeo, a party, and an obstacle course. The competitions involve speed records in which llamas must be caught, loaded with a pack, and led over a course that simulates difficulties typical of the trail, such as low tunnels, bridges, running water, and jumps.

TTEAM works very well for teaching llamas to respond quietly and well in all these areas. Getting a pack on a llama's back can often involve a struggle because most owners do not adequately prepare their llamas to accept the pressure of the girth. The llamas respond with fear or ticklishness, flinging themselves from side to side at the end of the rope. The following method works well for both llamas and horses with this problem.

Stroke the girth area with the wand first. Then use your flat hand and finally make Abalone circles first on top of the wool and then beneath it. When you see that the animal has calmed down, try Python Lifts on the girth area. After a few sessions you should be able to put the saddle or pack on without the dreaded feeling that you are about to enter a wrestling arena.

At one jamboree I saw a little girl in the holding pen with a young llama. She was about nine years old, dressed up in clean jeans and a vest with a rose appliquéd on the back. Something about the concentrated way she was talking to the llama reminded me of my own days with Highvale Architect Fairchild Dale.

"Seems like you have a pretty smart little llama there," I said. "What's his name?"

"Larry," she said proudly. "I think we're going to win a prize today."

I watched the way she put her arm around the llama's neck, and to my mind she had already won.

9

Animal Voices

The poem below, "Magic Words," is an Eskimo expression of a belief once shared by many cultures: that in long gone times humans and animals were one family.

In the very earliest times
when both people and animals lived on earth,
a person could become an animal if he wanted to
and an animal could become a human being.
Sometimes they were people
and sometimes animals
and there was no difference.
All spoke the same language
That was the time when words were like magic.
The human mind had mysterious powers.
A word spoken by chance
might have strange consequences.
It would suddenly come alive
and what people wanted to happen could happen—
all you had to do was say it.
Nobody could explain this:
That's the way it was.

—*Shaking the Pumpkin: Traditional Poetry
of the Indian North Americas*
edited by Jerome Rothenberg

HOW DO YOU TTOUCH A SNAIL?

When I was in Moscow, one of the things I especially noticed was how much the people love their parks. They use them for everything, kind of like great open-air community centers, or expanded outdoor living rooms. Even in the midst of winter, fur-hatted crowds congregate under the trees and along the rivers. People stop to talk, their breath steaming out in front of their faces in little puffs.

Therefore in 1985, when the Club Healthy Family (Klub Zdorovoy Semze), an association of four hundred or so family-oriented Muscovites, asked me to present my work to them, I wasn't surprised to find that the presentation hall was to be without walls or ceiling. I was to meet them in Gorky Park.

The name Gorky Park has long evoked sinister and melancholy images of murdered spies and bleak winter vistas, but I found something very different there that April afternoon. A mood of almost classical calm and grace prevailed on the banks of the Muscova River where an elegant nineteenth century garden descended in formal, stepped tiers to the water. The club members, a group of approximately one hundred fifty or so adults and children, sat quietly on blankets on the ground, their faces lit with the reflected sparkle of sun on water.

Through my interpreter, André Orlov, we talked about the TTouch for their animals and then had a practice session, with people pairing off to try the circles on each other, taking turns reacting as an animal would.

Afterward, as we sat and talked, a little boy raised his hand. "I would like to do the TTouch on a snail. How can I do that?" he asked.

Everyone laughed, but I thought it was an excellent question. "There's no problem," I told him. "You can touch the snail with your mind." He nodded, satisfied, accepting this new concept of communication as something he would try out for himself as soon as he found a cooperative snail.

I think most of us start out with minds as open and flexible as that little Russian boy's, but as we grow older it becomes more difficult to entertain ideas that are out of step with out cultural definitions of reality.

In our technologically oriented society we say we don't believe in "the supernatural," preferring to acknowledge as "natural" only that which can be explained logically. It's as though we are tuned in to a radio with a limitless number of stations, but we stay stuck listening to just one. We have trouble believing there are any other stations, even though we unconsciously sense their presence.

Native peoples around the world have always understood that all of life is interconnected in a vast network of interdependent being and have known how to tap into it for communication. To Native Americans or Australian aborigines, for instance, nothing is "supernatural" or beyond nature because nothing, including themselves, is apart from it. They think it perfectly "natural" to be able to hear in their own minds the thoughts of friends who are many miles away. In their seamless world they can hear stones speak, and ask plants and animals to share their wisdom.

How can we, without abandoning our rational mind, widen our awareness to be like that of the aborigine, to include so much more of the dazzling spectrum of information in which we live? How do we regain our inborn ability to hear with the third ear, to communicate with the sixth sense?

It's a very subtle process, really, different for everyone and difficult to describe, since by nature it's nonverbal in the first place. But happily, it's something that experience has shown us it is possible to learn.

We've discovered that practicing the TTouch gives you the opportunity to focus your awareness in a different way, a way that gives you access to a much wider range of communication than that of "ordinary" reality. When I first heard a snake actually "talking" to me I was afraid to tell anyone. I thought everybody would think I was crazy. As a matter of fact, I even entertained that thought myself.

But over the years, as I continued to work with animals, I realized that I was truly sensing their feelings, and that the TTouch is a key that can help to open the inner ear not only to the voices of our own animals, but to the many voices of nature all around us.

In 1984, I initiated a series of studies of the brain waves of practitioners of the TTouch in an effort to understand more about how the method works. The research was conducted under the direction of Anna Wise, a specialist in biofeedback and brain-wave activity with eleven years of professional experience in England and the United States.

For our research we used the Mind Mirror, a machine constructed by Maxwell Cade. A renowned British psychologist, researcher in physics and biology, and author of *The Awakened Mind*, Dr. Cade invented the machine that measures and record the levels of beta, alpha, theta, and delta waves being produced in the brain.

Put very simply, as it functions the brain emits four frequency ranges or wave states. Beta waves, which are probably what most of you are emitting as you read this, represent the normal waking state for thinking, logic, and problem solving. Alpha waves reflect a type of relaxed, detached awareness that is like a bridge linking the conscious and the subconscious mind. Theta are the brain waves of the dreaming mind and also the waves of what I call the "aha" moment when you are hit with a sudden inspired realization or insight. Though delta are the waves of deep sleep for most people, healers and people who have very high levels of intuition and empathy exhibit them when awake.

As a matter of fact, when in states of concentration, highly creative people, people with psychic abilities, healers, and those who practice advanced meditation usually have all four waves operating at once.

We set up the experiment to see what the brain wave patterns of twelve students with only limited experience in TTouch training would reveal and Anna was amazed to find that when the students were actually doing the TTouch on horses, all those who had taken a minimum of one week's training produced all four waves. Those who had not had previous experience with the TTouch did not produce alpha, the connector between conscious and subconscious thought.

We continued the monitoring and found that the results were consistently repeatable, an exciting indication that the TTouch was capable of triggering a heightened level of perceptual sensitivity.

As you practice the TTouch and work with your animals, this "awakened mind state," as Maxwell Cade calls it, will become more familiar to you, and you'll find that it becomes easier and easier to enter it. Many people tell me how deeply the work has affected their lives, like a door opening into a whole new world. They talk about the delight of expanding their horizons, and of how much fuller and broader they feel as human beings. One woman said that learning to speak and hear the language of animals and nature was like experiencing the world change from black and white to color. Another student described his unfolding dialogue with nature as the sensation of a barrier coming

Stevi Johnson

Joyce, the Burmese python, and Linda in a moment of mutual appreciation. Many people are touched and amazed to discover that cold-blooded reptiles can have a warm heart.

Joyce coils back to observe the Raccoon TTouch.

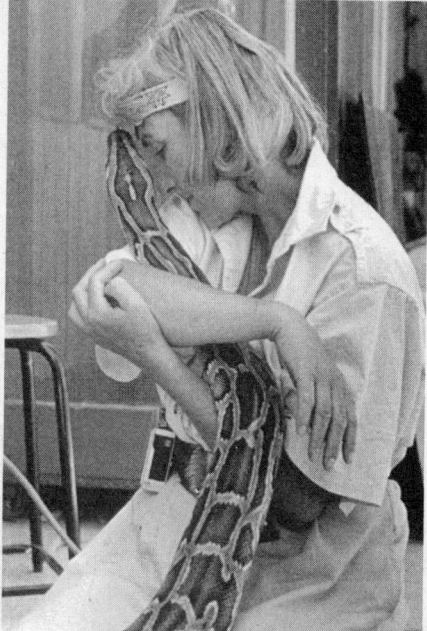

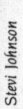

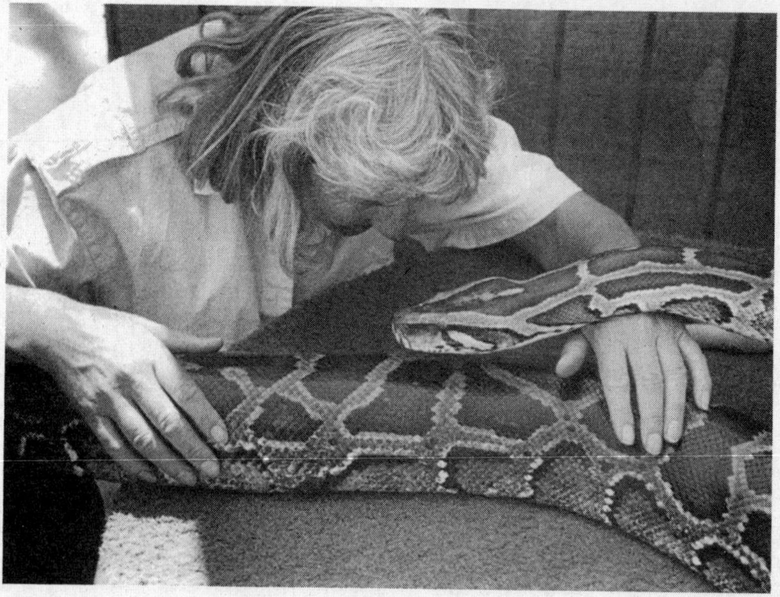

Stevi Johnson

Mr. Big Guy the iguana (see page 151) seems to be enjoying the TTouch for his arthritis.

A joyous reunion. Holey Fin the dolphin welcomes Linda back to Monkey Mia, Australia, after a two-year separation.

Stevi Johnson

Sylvia Rust

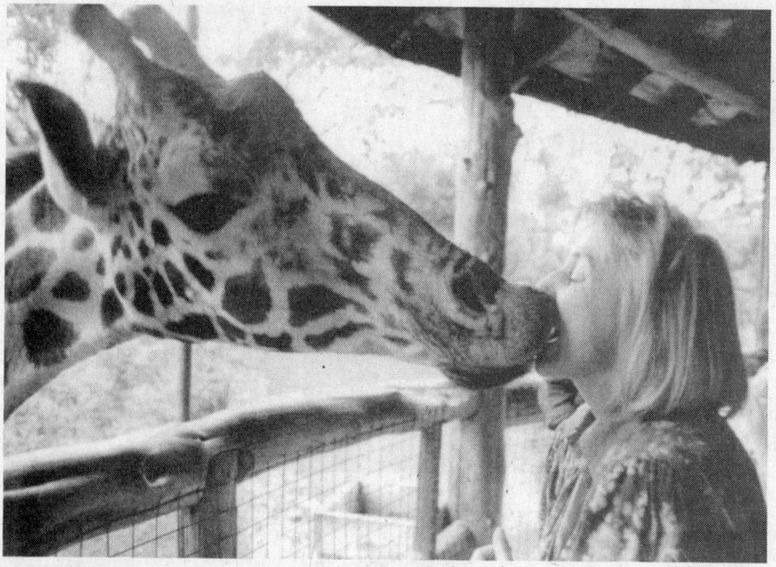

Jodi Frediani

What language is this? Linda and friend in Nairobi.

Liz Turnquist

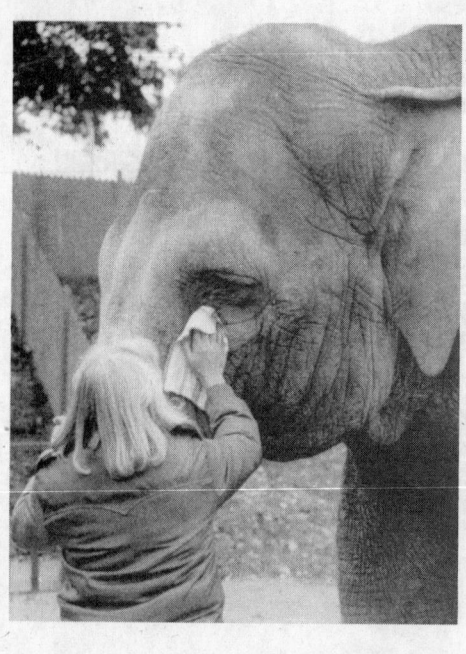

To gain elephant Shanthi's trust, Linda began work around the eye area, where Shanthi was comfortable about being touched. Twenty minutes later, Linda was able to TTouch the elephant anywhere on her body. Using a warm, damp cloth to make circles is especially soothing.

Practicing the Tarantula Pulling the Plow TTouch at the Chimfunshi Wildlife Orphanage in Zambia (see page 229). Left to right: Tracy, Linda, Sandy, and Harriet.

Tober, shown here riding "chimpback" on Linda, is one of the orphaned primates living at the Chimfunshi sanctuary. Every morning the chimps are taken out for a day of adventure and play in the African forest.

Harriet Crosby

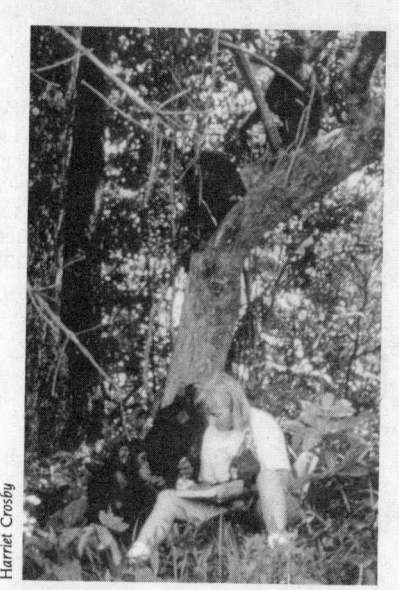

Linda at Chimfunshi doing research on the book with a special advisory committee.

"This feels good" is the message Linda receives from Boo Boo—the chimp was suffering from a slight fever. Ear work helps to reduce fever and gives an overall boost to the body's regulatory systems.

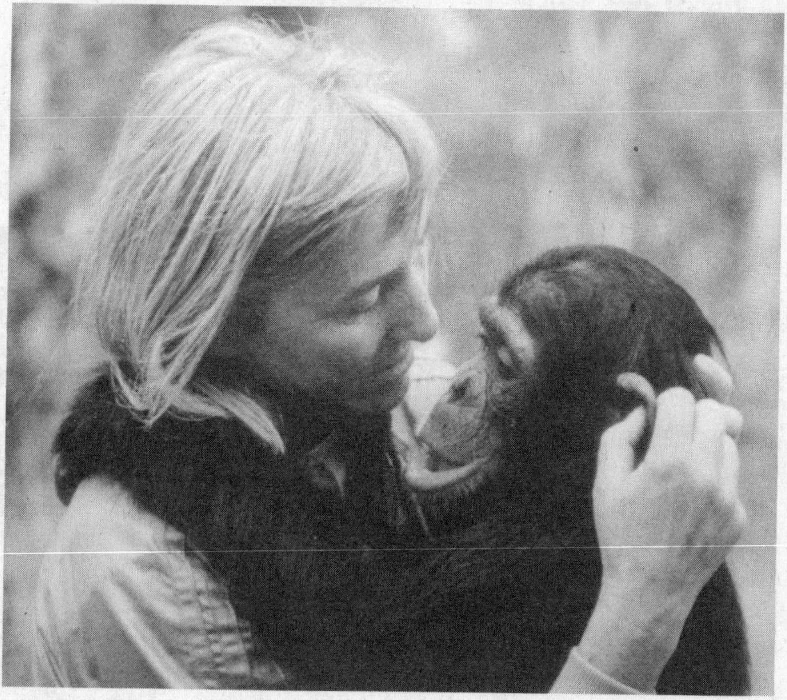

Harriet Crosby

Reaching across the language barrier: Linda and five-month-old Kenyon, a black bear cub orphaned by the Montana forest fires of 1988 (see page 222).

During Kenyon's rehabilitation in Texas, a washtub full of water was his favorite hot-weather substitute for the freshness of a Montana mountain stream.

A rescued victim of suburban overpopulation, this red fox now makes her home in the Wildlife Rescue and Rehabilitation Sanctuary, in Texas (see page 217).

Copper Love

Linda is using tiny Raccoon circles on the mouth of this little ferret, who was a biter. As a rule, two to three sessions on the mouth, body, and hindquarters will stop the biting habit.

Stevi Johnson

Courtesy of Dr. Ewald Isenbugel

Ewald Isenbugel, veterinary director of the Zurich Zoo, prepares a snow leopard to receive a shot. Linda worked with the leopards (see page 40), who come willingly to the fence to visit, and even more astoundingly, to receive medication.

Linda uses the wands like an extension of her arms to make the initial approach to two young feral cats (see page 47).

Kiki

down, a barrier that he himself had created by imagining it there in the first place.

It's been my great pleasure to find out myself that as this imaginary barrier melts, communications and encounters are possible that transcend normal perception of language, time, and space.

IN THE TIME OF DOLPHINS

A few years ago in Australia, I had such an encounter with a dolphin. Practically everyone has heard both ancient and modern tales of the curious and magical affinity between dolphins and humans, stories of dolphins rescuing swimmers from drowning, developing friendships with people, taking children for joyrides.

Many indigenous peoples give dolphins a uniquely important place in their cosmologies. The Native American Hopi, the African Dogon, and the Australian aboriginal people all share the common legend that dolphins originally came to earth from the stars and have a gift of great wisdom to offer humanity.

I met Holey Fin, my dolphin friend, in Australia in 1982. I had gone there originally to attend an international conference concerned with environmental issues and then, spellbound by the land, found myself staying on and on . . . and on. At the conference I encountered several people who felt instantly familiar to me, the way it sometimes is with new friends, and we took off together on a journey into the outback.

Australia has an odd, magical intensity unlike any other place I've ever been, as though it is still charged and echoing with the primordial forces that shaped it. This sense of primal presence is even more concentrated at certain spots, special places like Ayers Rock, a lone sandstone outcropping that changes color with the changing light in the sky. Ayers Rock is the world's largest monolith, one thousand feet tall and thirteen kilometers around the base. Its origins are still a mystery. The Australian aborigines call it Uluru and say it is the center of the universe.

We first saw Uluru at twilight, rising like a glowing, orange beacon from the surrounding flatland. After dark, when the tourists had all left, we climbed its flanks to spend the night beneath a field of stars so bright they lit our faces and infused our dreams.

At the end of a week's traveling, we sat down one evening with a map to decide where to go next. Suddenly, my eye was caught by the symbol of a blue dolphin swimming in an inlet labeled Sharks' Cove. The cove appeared to be on the west coast in an isolated, almost roadless area of the country. Perth, the nearest city, was about nine hundred kilometers south. On the shores of the cove was a village called Monkey Mia. I checked the map's symbol key and read that wild dolphins visit Monkey Mia ten months of the year.

There was, of course, no longer the slightest question of where we were headed. The rental car people gave us a funny look when we told them our destination. "It's pretty much off the beaten track," the man said. "We only insure the car as far as Geraldton."

Uh-oh, I thought. Geraldton was only half way there.

The road was narrow and the traffic almost nonexistent except for the caravans of cattle trucks that would go roaring past us like boxcars. After a night and a day driving through desert, we reached the ocean, passing through the tiny town of Denim with its one hotel, built of beautiful, shiny adobe made of ground-up shells. Fifteen kilometers farther on was Monkey Mia, not a town at all but a trailer park that consisted of about thirty caravans parked right on the sandy beach of Sharks' Cove. The view down the coastline continued pure and almost uninhabited for nine hundred kilometers south to Perth. There was a weather-beaten clapboard store selling provisions like Spam and beans and flashlights, and a big communal wash house with seawater showers.

The friendly couple who ran the place rented us a campsite. She was plump and motherly with the freckled complexion of a redhead. He was smaller, with thinning hair and a leathery face. He smoked a lot of cigarettes, holding them in the corner of his mouth and squinting against the smoke.

Why was the inlet called Sharks' Cove? I asked them nervously. Were there any? There are only two animals in the world that really terrify me: sharks and leeches. No, we were assured, there had been once, but the dolphins had chased them away.

The pod had first began coming to the cove twenty years earlier and had returned every year since, staying ten months out of the year and thoroughly enjoying their contact with humans. Visitors would constantly feed them frozen fish, which they would politely accept. Often they would swim off a little way and discreetly spit the fish out, obviously much more interested in the visitors than their gifts.

There were about twelve dolphins in the pod, but they came in to visit in groups of four and five. All had been given names based on their different characteristics—there was Holey Fin, so named because of a bullet hole in her fin; there was Nicky with a notched fin; Goldy, who still had the golden coat of a young calf; and Joy, whose whimsical playfulness was just that.

The dolphins were friendly to adults, but it was the kids they really loved. As soon as children entered the water, three or four of the dolphins would swim over to play. The children would stand in the warm, shallow water, clasp their hands behind their backs, and lean forward to kiss the dolphins on their "melons," the tops of their rounded heads.

The dolphins were a little shyer with grown-ups, swimming with them but usually staying just out of arm's reach, probably because in the shallow water adults loomed larger and were less intuitive in their approach. (I often saw people trying to grab the dolphins or stick a finger down into their blow holes). Obviously, the kids were doing something right, so I, too, bent down to present a low profile in the water and kept my hands behind me to offer a sleek, nonthreatening body image.

There was one adult, however, to whom the dolphins paid very special attention—an older woman with only one arm; the other had been amputated at the elbow. They would come right up to the woman and literally shower her with attention. She would make a hoop to her body with her "whole" arm, and the dolphins would swirl around and swim through it for her. It was beautiful to see how they singled her out for their most particular care and affection.

We stayed for ten timeless days. On the last morning I wanted to say good-bye to the dolphins but they were nowhere in sight. I stood on the shore and the water seemed very empty without them. We packed up the car and drove to the front of the store to fill up the gas tank at the pump. I turned for a last look at the sea and saw a streak moving along the shoreline toward us. Jumping out of the car, I ran to the water's edge. It was Holey Fin. She was alone.

We met in the shallow water that reached to my knees. Holey Fin and I had spent much time together. Normally she would just hang around and play with me. This time, however, as I bent toward her she not only let me kiss her on the top of her gleaming round head, but she reached up out of the water and returned the caress, touching my cheek with her beak—good-bye. (See photo insert.) Never had she done anything like that, not even with the kids. When I drove off, I was misty eyed.

Back home in California I pictured her and the other dolphins in my daily meditations, visualizing them at play and sending them my love and greetings. Two years later I returned to Monkey Mia for another week. As soon as I got there I put on my bathing suit and headed excitedly for the water. Sure enough, just as I had hoped, there was Holey Fin. She came toward me immediately, zipping along, a gray shadow in the clear turquoise water. I bent to greet her and she reached up and caressed me on the same cheek in the same way as before, only this time to say hello. In that instant the intervening two years completely disappeared. I am certain that she recognized me.

Once, when I was talking about this to an African medicine man, he told me a story about dolphins. His name is Eli Hiam and he is from the Upper Volta region of Africa. We were sitting on his blanket with a wide view across green hills to a cloud-dotted sky. Eli has the sort of eyes that look at you and seem to know everything about you and your life. He makes you feel as though there is nothing you would want to hide from him, even if you could.

Dolphins, he told me, have the ability to make their presence known to us even though we're not near the water. "Once you can see with different eyes and hear with different ears," he said, "then you will be able to look out across the land and see the dolphins rising up and coming through the sky to be with you."

"You see," he said, pointing out in front of us, "right there over the mountains is a whole pod coming up to visit us now."

I was fascinated because it was obvious that what he was seeing was as real to him as our own two bodies sitting there on the blanket among the conch shells that he had been showing me.

Sometimes now, when I look out at the sky across a valley or a desert and think of Holey Fin, an unmistakable sense of her presence comes to me, strong and clear as the light over Uluru.

THE POWER BEHIND WORDS

Communication with animals through the sixth sense can range from speechless mutual recognition, like my experience with Holey Fin, to a flow of words that appear spontaneously in your mind, like the snake's message to me. Sometimes I hear the words consciously and clearly. Other times they seem to float into my head and I just jot them down in a notebook like dictation, without being fully aware of their meaning. On reading them later I'm often astonished to find out what was said.

Animals don't chat with you directly in your native English, French, or Hindi. I don't know exactly how it happens but it's probably more as though an open, listening mind is a receiver for nonverbal consciousness the way a radio is a receiver for sound waves. The radio turns sound waves to language. A receptive mind translates nonverbal communication between animal and human into words and thoughts that can be grasped by your logical consciousness.

When your mind is calm, free, and open and your message clear, you can also become a sender. Perhaps this happens because animals hear and translate not the language in which you think and speak but the true meaning behind your words.

When this type of communication first started happening to me, I was wary of sharing it with others, but gradually I discovered that it was a far more common occurrence than is often acknowledged.

Michael Roads, one of the friends I had met at the One Earth Gathering in Australia, shyly confided to me one night that he had "conversations" with plants and animals. He wanted to tell people about it, he told me, but he was afraid of being labeled mad—or at best, sentimental and deluded. Michael's adventures and insights were so remarkable and vivid I encouraged him to publish them. By then I had noticed that more and more people were coming forward with similar experiences and were being received with growing interest and respect by the public. Talking with animals was no longer a secret that necessarily had to be kept "in the closet."

Emboldened, Michael wrote *Talking with Nature*, a book describing his transformative contacts with the natural world. In one of the chapters he tells of his early skepticism and how he learned to listen and let go. In another he describes his amazing encounters with a tribe of wallabies, kangaroo lookalikes who had overrun an upland meadow he had intended for his cattle.

Every night, whole families of wallabies emerged from the forest to gorge themselves on the delicious organic clover Michael had spent hours cultivating. Finally one night, in despair, he took his gun and drove out in his Land Rover, determined to shoot the wallabies and drive them off his land.

He sat in the car in the dark and waited. Hearing a rustle nearby, he switched on his high beam, grabbed his gun, and leaped from his seat. There, caught in the spotlight and brilliantly outlined against the darkness, was a large, three-foot, tawny-colored male. The wallaby stood upright on muscular haunches, his shapely, almost deerlike head turned toward Michael, eyes glowing red as jewels. A perfect shot. Michael lifted his gun and sighted.

But he couldn't pull the trigger. For a long moment man and animal remained frozen in position staring at each other. Then Michael quietly lowered his rifle and the wallaby leaped with one bound into the darkness.

In that moment of locked gazes something had happened to Michael—the spirit of the creature, its wild beauty, had entered and melted his heart. "Compassion, a comparative stranger to farmers saturated in death, surged powerfully from deep inside," he wrote later. "No more violence. There had to be a different way."

And there was. The next morning Michael stood in the middle of the pasture. Feeling ridiculous but determined, he called a message out loud to the wallabies, knowing their unseen eyes were watching from the surrounding forest. He told them that he wanted to make an agreement with them: he would stop shooting if they would limit their foraging to the edges of the field, leaving the center to the cattle. There was no answer of course, except an odd resonating quality to the silence around him. He left feeling self-conscious and a little foolish.

Next, Michael padlocked the gates to the pasture and told the neighbors there was to be no shooting on his land. Within a few months the field was growing so luxuriantly that Michael was able to increase the number of grazing cattle from thirty to ninety. The center portion of the pasture was dense with grass, and though it was crisscrossed with wallaby trails, not a blade had been nibbled. They had kept to the bargain, confining their nightly feasts to the perimeter of the field.

Unfortunately, the story has a sad ending. For three years the wallabies and the cattle peacefully shared the field, but at the end of that time Michael sold his farm and moved away. The new owners took the padlock off the gate and once more allowed shooting on the land. When Michael came back for a visit he was horrified to hear that the pasture had been decimated. As soon as the shooting began, troops of wallabies emerged in unprecedented numbers from the forest. Despite the carnage (six thousand shot in two years), they had swarmed over the pasture and literally wiped it out.

In spite of this sad finale, Michael never forgot that a miracle of communication had occurred. Nobody can explain such miracles, but one thing *is* clear—they happen.

ORCA SONGS

Michael has found an eager response to his books, and international attention and research are increasingly focusing on bridging the gap between humans and the rest of earth's creatures. One group that is deeply committed to such research is IC, or Interspecies Communication, a nonprofit organization that combines the efforts of biologists and artists. IC's founder, my longtime friend musician/ecologist Jim Nollan, has spent years communicating with whales through music. Some of his many contacts have been with orca, or killer, whales.

I was amazed to find that a mature male orca can weigh as much as eight tons and measure thirty feet long. Orcas belong to the class of whale called Odonteceti, Latin for "toothed" and "whale," a group that also includes dolphins and sperm whales. They have a single blowhole on top of their head for breathing, navigate their world by using sonar, or echolocation (whale sounds can travel as far as five hundred miles underwater), and weave songs so intricate and hauntingly beautiful they send shivers up your spine. I've even heard of people bursting into tears on hearing the songs for the first time.

In the summer of 1985 I was invited along on an IC research trip, part of an ongoing study known as the Orca Project. On this trip, the *Awesome*, a thirty-six-foot, spanking white research cruiser, was heading for a rendezvous with a pod of orca whales who were summering in the waters of Robbsens Byte, a strait between Vancouver Island and the mainland of Canada. IC had outfitted the Awesome with a state-of-the-art underwater sound system to broadcast, receive, and record the music, songs, sounds, and vocalizations that would be our language of exchange with the whales.

There were twenty-two people on the trip, including several children and a Tibetan lama, Tia Situ Rinpoche, and his monk assistant, who, while on a journey from their Himalayan home, had flown to Canada to join us. Aboard ship we waited eagerly for our underwater microphones to pick up the clicks and whistles and moanlike sounds that would let us know orcas had arrived. Sometimes Jim wouldn't wait, but sent his music out into the underwater world to entice the whales into our range.

Jim played electric guitars and synthesizers and the rest of us just used our voices. As soon as the whales appeared, flocking around the boat and leaping out of the water, our orca-human dialogue would start. We sang singly and in chorus. We whistled, imitated orca sounds, sang reggae rhythms, and made up arias and melodies on the spot. The orcas would pause and listen to us, then answer.

It was an endless multifaceted communication. Sometimes, when we changed our music or vocalizations, the whales would respond by changing theirs. Sometimes, when Jim played his guitar, the responses were so close to his music the result sounded like a single unbroken song. One particular orca vocalized with such a unique timbre and sounded so much like a bagpipe that we christened him Angus.

Music wasn't our only form of nonverbal communication. As a group we camped on an island in the channel, a haven where mosses grew and birds nested in the tall pines—a pristine home for bear and deer, toad and honeybee. Nurtured by the island, we grew closer to each other day by day. Gradually we seemed to be entering into a magical web of communication, a harmonious interplay between ourselves and the voices of the land, water, and sky worlds around us. Our inner ears became attuned to the unspoken thoughts of one another, to the speech of the orcas, and the words of the kingfisher and the golden eagle.

One night a few of us stayed ashore while the rest of the group boarded the Awesome. We sat quietly in a grove near the beach, waiting to see whether we could make further contact with the orcas, stilling our minds to receive whatever communication might come. After a while words completely disassociated from my normal thinking began forming in my mind. "We hear you, we hear you," the orca words began. I started to write in my notebook.

Suddenly there was a loud smack, the sound of a whale's flukes hitting the waves. We ran down to the shore and there directly in front of us was a tall triangular fin speeding toward us like a submarine. To our astonished eyes it seemed as though the whale was heading straight for the beach, but at the last moment, about fifteen feet from shore, he stopped with a motion so abrupt that water surged up onto the sand at our feet. Then he turned his enormous black-and-white body so that he was parallel to the shoreline.

In the beam of our flashlights we saw his knowing eye looking straight at us, while from the *Awesome* we heard the loud splash of excited whales leaping repeatedly out of the water. Very slowly, he then moved off, leading us along the coastline and keeping as close to us as he possibly could—a promenade for humans and whale, an interspecies walk in the beauty of the Canadian night.

The next night was moonless once more. The air was absolutely still and the normally ruffled water of the channel was glass smooth. There had been quite a bit of orca activity during the day so several of us decided to go out in kayaks and see if we could make contact once again.

There were six of us in four boats. Tia Situ and his companion, wearing saffron robes, were in one large kayak paddled by David Grace Charry. My friends Virginia (Gigi) Coyle and Joan Halifax were in the other two. In our little boats we skimmed across the sea, fragile as skeeter bugs, chillingly aware that an accident in these freezing waters would mean death in minutes.

The water glowed with phosphorescence and every stroke of the paddle stirred a trail of luminescent drops. The sky above, awash with stars, was perfectly mirrored in the still water below. Dipping our paddles into the reflected stars, we floated in a disorienting dimension somewhere between watery space and spaceful water.

Joan, an accomplished anthropologist and well-known Buddhist teacher, had traveled the world to many wild places, yet across the darkness I could feel her growing anxiety. Both as a child and an adult she had experienced periods of blindness and her journey into this boundaryless realm was an act of bravery. Gigi and I kept her kayak close between us with David and the Tibetans following behind.

After only about five minutes of paddling, the calm night exploded into a splashing roil of action. A pod of eight to ten orcas had surfaced all around us, leaping and blowing and putting on an incredible show. They stayed far enough away not to rock our kayaks, but my blood rushed and buzzed with excitement. In the darkness they sounded like a herd of stampeding, watery buffalo.

As the orcas wove around us, I began to feel an eerie physical sensation, a strong force as though the whales were bouncing their echolocating sonar off our boats and bodies. Every cell in my body vibrated with a sensation that was delicate and intense, like the beating of a hummingbird's wings.

Suddenly I heard a cry. One of the orcas was vocalizing above the water. The sound was like a high-pitched, piercing keening, so familiar it filled me with intense longing, yet so powerfully eerie that every hair on my body rose up. Like the howling song of wolves, the orca's cry goes straight to the heart, evoking an emotion that cries out in response.

I answered, and soon we were calling back and forth in the darkness. I felt that I was being directly spoken to. In the kayak behind me Tia Situ's assistant had begun to chant in Tibetan and I paddled on ahead, not wanting to break my connection to the orca. Joan was right beside me and Gigi behind. After about fifteen minutes the communication stopped. David paddled up and for some reason we stayed where we were, floating in a silent semicircle.

Then suddenly, in a gleaming rush, an enormous black-and-white body rose straight up from the water right in front of us. The gigantic dark shape seemed to tower ten feet above us, looking down, and then it slid carefully and soundlessly back into the smooth water, disappearing without a ripple. The orca had not allowed even the murmur of a swell to disturb the kayaks that rode the water, small as nutshells before him. Joan said it was like "seeing God."

Later, when we compared notes with Jim, who had been playing his music aboard the *Awesome* at the same time, he told us that the orcas around the boat had gone crazy vocalizing. Never had he heard any-thing like it, he said. He told me, too, that he had never heard of an orca vocalizing *above* water.

And that wild song still echoes in my ears, as does the music of the orcas and the poetry they gave me.

CLOSE TO HOME

Well, you might think after reading these stories, it's all very fine driving around the deserts of Australia and playing with orcas in the Canadian wilderness, but what does it have to do with me? I don't live like that.

Actually, if you live and work in a town or city, close communication with an animal can be even more important and meaningful than if you lived in the heart of a forest. Away from nature, or immersed in intellectual, indoor work, it's easy to develop artificial rhythms, to become unbalanced and lose touch with the vital cycles of the earth and the underlying harmonies that govern healthy life.

When I was living in Carmel, California, I was lucky enough to have as my caretaker a little black Scottie dog named Bonnie. Once in a while I would have to do a great deal of desk work such as answering letters and arranging clinics. Every evening the air would cool and a thrush would come to sing his bedtime song in the vines outside my window. For me, however, the waning day was simply a signal to switch on the desk lamp. Oblivious to my own fatigue, I went grinding on.

Bonnie, however, knew better. Observing me from her bed in the corner of the room, she would rise, curl her pink tongue in a yawn, and come over to tap me on the leg with her paw. The work day, she announced to her ignorant human friend, was definitely over. Then she would persist with these sly little taps until she finally succeeded in dragging me outside and down to the beach for the romp that both of us needed so badly.

A writer I know told me recently that while she loved her cat de-votedly, he was a real pest in the workroom. After an hour or so of quiet napping he would interrupt her at the computer, jumping up in her lap or stepping delicately across the top of the keyboard, asking for a change of pace. Had she picked up on his cue and taken a few minutes to stretch and play with him, she would not have risen from the computer with a stiff neck every night.

Even fish can communicate well-being. A study conducted by the University of Pennsylvania showed that patients about to undergo dental surgery calm down enormously if they are in a waiting room where they can observe a fish tank.

Quite literally, communication with animals brings us back to our senses. Go to a zoo, slow yourself down, and take time to establish a real contact with an animal. Do whatever feels appropriate. With a turtle or a snake or a fish you can slow your breathing and reach out with your mind. When I've done this with fish they often swim straight up to the glass where I'm standing and hover there looking at me. With a gorilla or an orangutan, playfulness is a marvelous language of exchange.

Once when visiting Art Goodrich at the San Diego Zoo, I went around to pay a call on the orangs. They were a family—mother, father, and youngster—and all three were very obviously bored, lolling around their grassy enclosure and ignoring the twenty or so people who stood watching them.

I had been shopping and was carrying a bag filled with my purchases. Very slowly I crouched down, opened the bag, and stuck my head into it in an exaggerated pantomime of looking inside. I did this a few times till mother orangutan suddenly noticed that something interesting was going on.

She moseyed over and leaned her long arms against the barrier, gazing spellbound as I slowly and deliberately reached into the bag and pulled out a dress. I held it up for her and she examined it with total fascination. Next I pulled out a T-shirt. She scrutinized this too, with the same look of intense concentration on her marvelously mobile face. Finally, I extracted a book. Tipping her head a little to the side, she seemed to be trying to figure out the meaning of the boxy thing I was now holding up.

Finally, I ran out of purchases. I don't know who was the sorriest— me, the orang, or the crowd of people who had suddenly been jarred out of their usual way of viewing animals.

Many zoos around the world are going through a spectacular transformation, from rows of concrete holding cells to beautifully landscaped habitats. Others remain dismal prisons, still waiting to catch up with the new era. But while beautiful enclosures are vast improvements over bleak cages, the fact remains that zoo animals are captives living out artificial lives in service to us. Many are depressed, bored, or stressed, and it's important to know that you can often make a difference simply by giving them your appreciative attention.

I'm constantly amazed at the levels of communication possible. At the Washington National Zoo an elephant named Shanthi had a surprise for everyone. I had been invited to come to the zoo to demonstrate the TTouch to a group of keepers. After my presentation, John Lehnhardt, the elephant collection manager, asked whether I would have a look at Shanthi, who occasionally suffered from flare-ups of arthritis, not uncommon even for an elephant as young as Shanthi. John was having a problem taking blood samples from her ear because the process made her nervous. Fear tends to lower blood pressure, which makes the veins withdraw from the skin's surface and become less accessible to the needle.

Elephants have an average life span of sixty years; Shanthi was a teenager. Born in Sri Lanka, she was an Asian elephant with a slightly smaller frame and smaller ears than African elephants have. She lived with her enclosure-mates Nancy, a thirty-four-year-old African elephant, and Ambika, thirty-nine and, like Shanthi, an Asian.

All three were waiting for us in their oval outdoor enclosure. It's amazing how small and fragile you feel when you're surrounded by three elephants milling freely. They tower over you, their massive bodies moving with a slow-motion power that seems unstoppable. You steer carefully clear of the huge feet, while fanning ears stir a breeze on your face. I always find it both remarkable and touching that such mammoth, slow bodies are endowed with such delicate feelings and such a quick intelligence.

John Lehnhardt had discovered his love for elephants by accident. Though trained as a biochemist, he was intrigued when he heard there were zookeeper jobs open at the Lincoln Park Zoo in Chicago. Did he have any experience with large animals? he was asked. A little, he said, with horses and cows. Great, they said, go and see Tony at the Elephant House. "That was fourteen years ago," John remembers. "I've been hooked on elephants ever since."

Shanthi was nervous at the beginning. "Steady girl," John said to her calmly, both to reassure and command her. "Steady there."

To reduce the elephant's fear of the needle, I had suggested that John do the TTouch on her ears, some days simply doing the TTouch without drawing a blood sample, other days doing it prior to using the needle. This would calm Shanthi and allow her to associate something pleasant with being touched on the ears.

With John to one side of her and an assistant keeper at her head, I reached up to one of her giant ears to demonstrate, making Lying Leopard circles all over its flat, palm leaf–like surface. After a moment the elephant reached around and began touching my face with the tip of her trunk. It felt as though a warm, slightly moist fingertip was delicately investigating my features.

"She's making friends with you," said the assistant keeper. Satisfied with my face, Shanthi then reached her trunk inside my shirt to explore under my arm. "Intimate friends," I said, laughing.

I continued the work on her, using a warm, moist cloth to make the circles (see photo insert). I've found that whenever there is fear in an area of the body, damp warmth is tranquilizing and comforting.

Then, reaching up once more, I made very light Raccoon circles around her eyes. I was exploring this area because she had no fear associations here, and I wanted to widen her pleasurable experience of the TTouch and deepen my contact with her. The skin was wrinkled and thick but took no more than a three pressure and her brown, almost human-looking eye with its long lashes had a concentrated, faraway look, as though she was listening to something.

After saying good-bye to Shanthi I wandered around the zoo. Somewhere on the path between the elephants and the lions I encountered the assistant again. "Oh, I was wanting to talk to you," she said excitedly. "Wasn't it great what Shanthi did with her eye?"

"I didn't know she did anything special," I said.

"You didn't see it? You probably couldn't from the angle you were standing. After a few minutes of you doing the TTouch on her one eye, she lifted her trunk and tried to do the exact same thing on the other eye herself. It was so amazing, I could hardly believe it."

We stood a while marveling at Shanthi's response, but in thinking about it later, it dawned on me that the behavior we considered so amazing was probably nothing so special to Shanthi herself. She was just being Shanthi.

The real marvel, I realized then, lay somewhere in between, not in the mind of elephant or human but in the wordless exchange between them. What can we call this exchange? Is there a name for what you experience? I guess *communication* will have to do.

But why not try to touch a snail and see?

10

Saying Good-bye: Easing the Transition into Death

Sally Forth

Your death
paralyzed the poetry in me
as surely as the pentothol
paralyzed
first your limbs, your lungs, and
finally your intrepid pony heart.

It was time: you'd grown old
beyond eating, old beyond rising
should you lie down as horses do,
confident of life. Your eyes
had dulled, your hide had no
resilience though you still
arched your neck to guide
my hand, still laid your muzzle,
texture of newborn puppies, on
my arm as I led you down to
the pasture. I assume you
knew, knowing me, and came

as businesslike to death as you
had to thirty years of life.
Two weeks have intervened and still

each night I hold you for the needle,
feel you slump, gravitate, hoofs
exposed like four loosed moons, watch
the earth moved, watch the body,
dragged beyond my hand, miraculously
fall with muzzle resting on
curled under limbs. You lay as you'd
lain, young and alive. Watch
the earth blanket you, understanding
then that she rocks more to rhyme
and rhythm than ever poets do.

—Marion Copeland
TTEAM member

CELEBRATION AND MOURNING:
THE FULL CIRCLE

The young blue jay sat dreamily in my hand, light as a puff of blue and gray dandelion down. His head was tucked low into the fluffed-up feathers of his body and his eyes were closed. When he opened them, black pupils thinly encircled in gold reflected the dazzle of morning sunlight. He was very still, but in this stillness there was also motion, a patient and quiet passing into death.

The night before had been less quiet. I was having a small dinner party and was just serving dessert when I heard a ruckus in the bushes under my dining room window. Leaving my guests to their raspberry tarts I went outside on what I was pretty certain would turn out to be a rescue mission. Sure enough, my two cats, out on their evening prowl, had cornered a young blue jay and had just begun to "play" with him; his left wing hung limp. Grasping the bird around the wings, I picked him up and spirited him inside, leaving two disappointed feline hunters wondering at the weirdness of humans.

The bird was stiff with shock, nature's merciful tranquilizer, so I went in under his feathers, feeling the matchstick delicacy of his body with one finger, working on his head, chest, and ear orifice with tiny

Raccoon circles (see page 244). Once again I was amazed at the speed with which the TTouch will revive a bird in shock while taming him at the same time.

In minutes I was able to take the jay into the dining room to sit with me, and there he stayed, exhausted but relaxed, tame under my circling finger and seeming to take comfort in my hand. Neither he nor I had ever had a close encounter of the blue jay–human kind and we spent some companionable time together before I settled him in a box of leaves and twigs for the night.

I looked down at him as he cozied himself into a corner of the box. He had lost some of his tail feathers in the skirmish, and the damaged wing, which I had been working on, still hung limp. His eyes were alert and his respiration and heartbeat were regular. It's tricky, I thought, small birds often don't make it through injury. The fragility of their bodies is made for flight, not fight.

On the afternoon of the next day he died—one moment the light of life still shone in the depths of his eyes, then softly it dimmed, and finally it was gone. Holding him, I watched it happen so gradually, so easily, I thought of shafts of sunlight on a cloudy day, the way they dim and disappear from one spot, only to reappear, brilliant as ever, in another. I saw in the passing of his life a motion from one form to another, endless as the cycling of seasons, an autumn leaf falling to earth to nourish the trees of spring.

Later, I looked out of the window at the cats basking on the sun-warmed flagstones in front of the house. The fluffball white Persian was daintily cleaning between the sharp claws of her outstretched foot. There would be other birds for her to hunt, I thought, and some would live to fly away and others would not.

The night before I had heard coyotes hunting, calling to each other from the arroyos and mesas surrounding my adobe house. Their wild yipping informed me that the cats could no longer be safely left outside for the night, yet in the crisp darkness of the New Mexican midnight it was a beautiful song to hear. Untamed and jubilant, it spoke to me of hunter and hunted and a continuum as circular as the wreaths of fresh flowers with which we commemorate the dead.

Animals teach us about death, and we in turn can help our own animals in their dying by knowing both how to support them and how to let go when the time comes. Often, we just can't face the sadness of saying good-bye to a beloved pet; we fear the moment of separation and the empty place that pet will leave behind in our hearts and in our homes. Too, we often feel guilty. Even though a pet is mortally ill or suffering from old age, making the decision to allow a dear friend to die is a hard one as long as the animal still seems to be hanging on.

Ironically, an animal often lingers on only out of attachment to us. Mirroring our fear, feeling our confusion and unwillingness to part, the animal will loyally continue to drag along, suffering pain, disorienting medication, and indignity for our sake, when release would be the natural and most welcome conclusion.

That's what was happening to Jake. Jake was a black Labrador who had been with his family for fifteen years. When I first met him he was in sad shape. I was leading a clinic at an Arabian stud farm in Aptos, California. Twenty of us would be sitting in a circle on the grass, when along would come poor Jake to plop himself down with us. Everyone would groan.

The dog was incontinent, had an incurable ear infection, and was so allergic to fleas that he had scratched all the hair off his bottom. As if this wasn't enough, his deteriorating condition made him smell so terrible that when he came around, desperate for love and attention, most people would feel compelled to either ignore him or shoo him away.

His owners, too, turned a blind eye to his condition, not from cruelty or lack of love but because their very caring made them want to escape their feelings of helplessness. After several days of watching this agonizing situation, I decided to risk sticking my neck out. Had they thought of putting the dog to sleep, I asked, trying to phrase my question as delicately as possible.

No, they wouldn't hear of it, they couldn't face doing that. They were adamant.

But the next day the subject came up again, and I pointed out the incurability of Jake's unhappy condition, suggesting that we ask the vet to come to the house as so many vets now do. That way, Jake could die at home in familiar surroundings and free of fear. I suggested, too, that before the vet came we have a farewell party for the dog. We'd sit with him, groom him, wash his body in soothing conditioner, and bathe his spirit in love and appreciation. We would turn his passing from a frightening and wrenching parting into a celebration in honor of his life.

The owners said that they would consider it, but in any case they would rather do it alone, and the subject was dropped. When the four-day clinic was over, I left, feeling sad for miserable old Jake who came hobbling over to my car to say good-bye.

The next thing I heard was that Jake's owners had given him his farewell party after all and had buried him at the big iron farm-gate where he had spent so many happy years as self-appointed meeter, greeter, and canine chief of security. Burying him there with a little ceremony gave his family a focus for mourning and the sense of completion that we all need when faced with a death, whether human or animal.

I never realized just how necessary such acts of respect and mourning really are until the death of Empress the elephant. In 1986, when I met her at the Honolulu Zoo, Empress was fifty-four years old, very popular with zoo visitors, and famous as the oldest elephant in captivity in the world. I had been asked to come to the Honolulu Zoo in Hawaii to demonstrate the TTouch for the zoo's keepers and to see if I could help the old elephant. After many years of standing on concrete day in and day out, she had developed severe abscesses on all four feet.

When I first saw Empress, she was leaning her trunk on the concrete retaining wall, trying to take the weight off her feet. She was obviously in pain, and I was amazed and horrified to see tears running down her magnificent face.

Pain to that degree makes many animals ferocious, but when I went into her enclosure, Empress was not only gentle but extraordinarily considerate. Elephant-style, she got acquainted by investigating me with the warm, moist tip of her trunk, touching my face and body with the care of a mother with a newborn infant.

I worked the Python Lift on all four of her massive legs and the Clouded Leopard on the bottoms of her feet. The abscesses were very bad; at times, the keepers told me, they had even exuded bits of bone, but the aged elephant stood patiently. While it was clear to me that she appreciated the TTouch and the attention, I sensed that she really did not want to live any longer and would be happy to pass on to "elephant heaven."

Her devoted keepers had tried many ways to alleviate her pain: they had used drugs, boots, and compresses; they had taken her out of her enclosure for relief periods on the cool grass under a giant shade tree. But it was all too late: the years on concrete had taken a terrible and irreversible toll.

There was no soft place in her enclosure for her to lie down, no soft dirt or sand to blow all over her body with her trunk the way elephants love to do. The "sand" that had been recently added to the enclosure was actually bits of volcanic rock that had been piled in the center and scattered about on the concrete. The particles were large and sharp-edged and hurt her to stand and walk on.

That night, after I had worked on her, Empress came to me in a dream, asking for release from service. She loved all the people who had come to visit her over the years, but the pain was simply too great to continue. I awoke to the sense of her presence and lay there visualizing a peaceful leave-taking and a memorial service to honor her passing.

Only a part of my dream was realized. The next day, on my way out of the zoo to catch a plane, I saw Empress standing in the shade under her big tree. Several of the keepers who had been her friends stood close beside her and a photographer was taking their picture. It was to be my last sight of her.

A few hours later she received her wish for release. The zoo made the decision to euthanize her. The time had come. I was told by a zookeeper that in the evening, after the zoo had closed and she was being led to the veterinarian's clinic, she lay down on the ground herself with no prompting and waited quietly, apparently knowing exactly what was about to happen. Shortly after that she was given an injection, and in moments she was gone.

The next day a Honolulu newspaper reported not only her death but also the fact that her body had been trucked away during the night and dumped into a landfill. The decision to do this was a practical one, I know. The zoo director had a pressing need to dispose of the body of a very large creature, and the very human desire to get something sad over and done with as quickly as possible.

The trouble with this ending, however, was that it robbed the old elephant's life of the dignity, love, and respect it deserved. It left the many hundreds of people who had known and cared for her with no way to express that respect, and nothing to mark her life and loss except the sad and degrading image of a dump.

In his book about the Philadelphia Zoo, *The Peaceable Kingdom*, John Sedgwick tells of the effect of the death of Massa, a fifty-four-year-old gorilla beloved by public and zoo personnel alike for his charismatic intelligence.

> When word of Massa's death went out across the country and around the world, the zoo received several hundred letters of condolence from individuals, ranging from the keepers of the gorilla house in Tokyo's Veno Zoological Gardens to the nine-year-old in Philadelphia who sent a tiny silk flower in a harmonica box. The staffers also handled a number of inquiries about where Massa's funeral service was to be held and where he would be buried. Responding personally to each letter, Debbie Derrickson [zoo public relations] and her assistant merely thanked the letter writers for their sympathy. They did not say that Massa, being a gorilla, would receive no service and no burial.

Though Massa's body was given to researchers for study, the zoo administration honored him and the special affection he inspired with a statue. Seeing it for the first time, John Sedgwick wrote,

> When I arrived in Philadelphia in July 1985, a sculpture of Massa by Eric Berg stood surrounded by piles of brick, stacks of lumber, and other materials waiting to be assembled into the new World of Primates exhibit, a splendid open-air habitat that would go up on the very site of the increasingly dreary Monkey House where Massa had lived for so long.
>
> It was odd to see Massa there in bronze, staring off into space, oblivious of all the junk around him. But little by little, as the months went by, the piles diminished, the new exhibit took shape and Massa's form emerged from the

rubble. It was an omen. Like some patron saint, Massa was an inspiration to the zoo staff. His body may have been scattered about the country (to various research institutes), but his spirit lived on at the Philadelphia Zoo.

Death makes no distinction between humans and animals, and neither does love. Why shouldn't we celebrate, mourn, and remember our animal friends as we do our human ones, with a ceremony that takes joy in the life that was, and that allows us to grieve and also to heal our grief?

RELEASING YOUR ANIMAL

Many people have great difficulty confronting death. Often patients die alone in impersonal hospital rooms because family members feel too overwhelmed and inadequate to go through the transition with them.

In the same way, many people are too afraid to stay with a beloved pet in its dying moments. Instead, they leave their dog or cat or horse to be put down by a vet in a clinic, and there, no matter how kind the vet is, the animal dies alone, a stranger in a place that is full of the smell of fear.

Veterinarians, aware of this, are increasingly encouraging people to see their pets through the experience of death. Over the past few years I've been running into more and more doctors who compassionately refuse to put an animal down unless the owner is present.

Many years ago, my husband Went repeated to me a conversation he had had with a Mormon deacon. The deacon's insight on the death process made such a big impression on me that I've been passing it along to others ever since.

"Suppose," the deacon told Went, "that you are a fetus. Someone tells you that one day, come what may, you are going to be thrown out of your womb home and pushed and pulled and squeezed through a narrow corridor at the end of which you will pop out into a world you can't possibly even imagine. There you will be held up and given a smack till you cry and then your life cord will be cut. The whole process, you are told, is called birth.

" 'No,' the fetus would cry, 'how is that possible? I will have no life support. Where do I get my nourishment and protection? I don't know what's through that narrow corridor or what's going to happen to me when I get there. It's terrifying. I don't want to go.' "

The deacon smiled at Went. "How do we know," he said, "that the process we call birth is so very different from the process we call death?"

When we think of ourselves as midwives helping in the process of dying, it makes the going that much easier. Communicate your feelings to your pet. When you do—crying if you want to, but bringing the good times you had together joyously to mind, too—your animal will feel the strength of your friendship and will know that you will never forget him or her. Mourn and celebrate at the same time, and don't try to cut the emotional bonds.

At this time the nonverbal connection of the TTouch can be especially useful. It offers you something practical to do instead of just standing by and feeling terrible, a way to stay in close contact and express your feelings, an active way to ease the journey of transition. Using the TTouch will comfort and calm both you and your animal, enabling emotions to be strong but not out of control. And that's important, because when animals sense hysteria in their owners they become upset themselves, and have a much harder time making the break.

In the right circumstances most animals will have no trouble leaving and some will even lose their pain and depression as soon as they realize death is near. The story of my friend and neighbor, Anne Milliken, and her Arabian mare is a case in point.

One day at feeding time Anne noticed that Estrella was standing on only three legs. She thought the horse had probably bowed a tendon running in the pasture and called the vet. His verdict after an X-ray—a shattered pastern and fractured coffin bone. His prognosis—less than a 10 percent chance of satisfactory healing, and if the bone did finally knit, a lifetime of pain and discomfort for the horse.

The doctor was sympathetic but could offer no hope. Just the summer before, he told Anne, he had cast a similar break on a much-loved junior roping horse. After two months and a great deal of suffering for both human and horse, they were forced to give up. He suggested that

Anne seek a second opinion, but his recommendation was to put Estrella down.

Anne was torn apart. She dearly loved Estrella and rode her every day. Together they had explored the secrets of the New Mexican desert, trotting in shaded arroyos where the red clay walls were pocked with the burrows of birds and lizards. Together they had worked hard to master the art of dressage, learning to move in harmony as partners in a dance. Besides, as a highly trained professional acupuncturist and TTEAM practitioner, Anne found it doubly hard to accept that the best course of healing for her horse was death.

Distraught, she phoned me and we met to discuss the situation. We called Dr. Matthew Mackey-Smith, a friend and veterinary adviser to *Equus* magazine, for a second opinion. He, too, believed that the three broken bones in Estrella's leg presented a bleak outlook at best and recommended putting her down as the most merciful solution. Though filled with grief, Anne knew that the time had come to say good-bye.

We called a small group of friends for a farewell party and Anne drove on ahead to the veterinary hospital. Ever since her injury, the mare had been depressed and in pain, but when we arrived we were amazed to find her out of her stall and having a wonderful time visiting with the other horses and nuzzling a pet pig.

I suggested that we approach her death as the Native Americans might have, holding in our minds an image of the injured horse completely recovered and galloping free to heavenly pastures. By projecting this image together, we would be united in supporting Anne in her grief when the mare was put down.

We then worked with the TTouch on Estrella's body. She especially seemed to appreciate the work on her ears, and the long comforting movements of the Lick of the Cow's Tongue (see page 250). After a while, Anne took the horse off to be alone with her for a few minutes. She sat on a bale of hay, watching Estrella graze on the early spring grass, green against the patchy snow, and recalling the magical hours they had spent together.

The vet had considerately told Anne to call him when she was ready, and she gave him the nod just as the evening sun slid down below the horizon. We took the mare to a grove of trees, fed her a little grain and some carrots, and worked on her for a few last moments of contact. Then the injection was given and within seconds the horse was down and the heartbeat gone.

After sitting quietly beside her mare for a few moments, Anne took a pair of scissors out of her bag and cut a keepsake lock of the black mane that still fluttered in the wind. Then, as we walked away from the bay body lying in the snow, we looked up and could scarcely believe what we saw. Perfectly outlined in the sunset sky above the roof of the hospital was a giant, peach-colored cloud in the shape of a horse's head and neck in racing position, neck stretched forward, ears flattened in the wind, nostrils flaring. We stared in silence as it streamed past us through the sky.

That night over dinner we drank a champagne toast to Estrella from silver-tipped sheep horns, drinking vessels that are still used by mounted sheepherders in the mountains of southern Russia. What could have been a lonely and sad day for Anne turned out to be a gift to all of us.

There is a postscript to this story. Several weeks later I found myself at the Seattle airport with a few hours of free time between planes. Thinking it might be a great way to snatch a visit with my Seattle friend Viki Menear, I phoned her to see if she could jump into her car and come for a chat on the wing. She arrived half an hour later with a hug, and a parcel that looked like a stick carefully wrapped in tissue paper.

"I'm not sure why I brought this," she said, handing it to me. "I just grabbed it as I was going out the door. It's for someone whose favorite horse has just died and I thought you might know such a person."

I was stunned. It was an Indian-inspired horse effigy, a "spirit of the horse stick," about twelve inches long with a horse's head at one end and a perfectly shaped hoof on the other. The carved head with two tiny feathers braided into the mane was thrust eagerly forward in a racing position, as though the horse was galloping full tilt into the wind like the cloud racer we had seen in the sky.

Viki told me that the stick had been carved by Nickki Niemann-Lee. She and her husband, Greg, create figures inspired by Native American culture. In the tissue paper wrapping I found a little card explaining the effigy's background:

With the advent of the horse there came a great period of prosperity for the plains Indians. Hunting buffalo became a much easier task as did many other things. Because of this the horse was much revered by the Indian people. A warrior would paint his horse with the same patterns that he used on himself. These patterns had as much to say about the quality of the horse as they did about the exploits of its owner. Often when a horse died bravely in battle or in a raid an effigy was made in its honor.

I brought the spirit stick home to Anne as a tribute not only to the spirit of her friendship with Estrella but to the undying bond between all animals and humans.

CHILDREN: SAYING GOOD-BYE TO PETS

The other day I was disturbed to hear an acquaintance say he thought it was bad to let children raise pets because it dooms them to failure.

"Dooms them to failure?" I asked, astonished.

"Yes," he said. "Given the short life span of most pets, children will inevitably have to watch their animals die."

I told him I didn't agree with him at all. It seems to me that living with an animal and experiencing the full cycle of its life is a great teaching for a child, a way for him or her to empathetically witness the phases and passages of life that all of us must go through, and a means to learn about them without fear. Too often, in our technolog-ical urban culture, children experience death as isolated from the life process, something sudden and violent on television, or hidden fear-fully away behind the closed doors of Intensive Care.

Two years ago I had an exceptional opportunity to see for myself just what a positive thing it can be for children to be present at the death of an animal they love. I had arranged for a week of camping for forty children between the ages of nine and sixteen, students of the international Nijoni School of Global Consciousness based in Santa Fe, New Mexico. Most of the children had already received beginners' instruction in the TTouch and were now ready and eager to work with more difficult animals.

North of Santa Fe, where the Sangre de Cristo mountains rise pine-green and streaming with rivers, I found a private preserve and summer lodge in the midst of a national forest. There was a fine main house and dining room and, best of all for our purposes, the property was leased by a woman who raised crossbred wolf dogs and had about thirty of them on hand. Because she did not have the help or money to train the dogs, she kept most of them tied to trees or in pens.

Wolf dogs can be difficult animals to have as pets, and I believe that breeding them is, in most cases, a bad idea. More often than not, they are one-person creatures with a reputation for roaming, which means that they must be kept chained or penned. Still, I felt that this situation was an excellent learning opportunity for the children— where else were we going to find a large group of animals, a person who wanted help with them, and a perfect place to camp, all rolled into one?

So off we went to the mountains for a week. The woman kept her puppies and dogs under three years of age in pens. They were so wild that I had to go into the pen with wands in order to calm them down enough to catch them. Once out, the dogs were so terrified they went into freeze reactions (page 111). To get them to move, the children had to work the TTouch on them, sometimes for as long as twenty or thirty minutes. Taming them to lead took a number of different tech-niques. Two or three children would work on one animal together— doing the TTouch, offering food, stroking with wands, and finally work-ing with the Journey of the Homing Pigeon (page 272) and the labyrinth.

Initially, some of the dogs were so tense they wouldn't eat until the children had worked on their mouths for a time.

Day by day I watched the children grow closer and closer to the animals—it was wonderful to see them learning how to be careful and patient. Some of them were working on a very old female, the mother of many of the younger dogs. She was painfully thin, her fur was sparse, and she was disoriented and wobbly. I thought she might have a neurological dysfunction of some sort.

As a matter of fact, after observing her, I felt that she was ready to go, and was only hanging on for the sake of her owner, who loved her very much. I began hoping that she would die before we left, so that we would be there to support both dog and owner when the moment did finally come.

One day, toward the end of our week-long stay, she lay down and remained motionless for many hours. The next day, she got up (she was one of the few dogs who had the run of the place), made her way to the puppies' pen, and collapsed there.

Eight or nine children who had worked on her and grown to love her gathered around. They put their hands softly on her body, and took turns doing the TTouch. After a while, with the children encircling her, she let go and life passed quietly out of her. A moment later, the air was full of the howling of wolf dogs mourning their loss.

I remember one twelve-year-old boy who had never lost a pet before, talking about his mixed emotions. He felt like crying, he said, and at the same time he was happy that he was able to be there for the dog and help her to go, particularly since she had spent so many months hanging on. She was physically gone, he said, but he knew he would never forget her.

FEAR OF THE DYING

Many people find it hard to acknowledge that they are afraid of a dying person or animal, and apprehensive about touching them. Admitting to such a fear makes them feel ashamed or disloyal, but it's a very understandable and primordial reaction. Dying is often neither clean nor pretty, and the living can feel deeply conflicting impulses—to give comfort on the one hand and to avoid contact on the other.

In cases like this, the TTouch can be extremely helpful, as my collaborator Sybil Taylor found out through personal experience. Sybil had gone to visit her friend Evelyn, who lives in a small New York apartment building, a mini-community where all the tenants know and care about each other like an extended family. Evelyn's downstairs neighbor Robin, a motherly woman who never could resist a stray, whether four-legged or biped, had just died, leaving behind two parakeets, a faltering little Maltese dogette, and an old tortoiseshell cat named Gregory.

The animals reacted to their bereavement physically. They behaved in confused ways and the dog and cat both fell ill. So the members of the building's "family" banded together and adopted Robin's lost brood on a care-share basis, giving them the attention and love they needed until permanent homes could be found for each of them.

Gregory had been the woman's first favorite cat, the one she was most worried about leaving behind. A character known to everyone in the building, he was a social lion, headquartering himself on the landings and stairs to meet and greet arrivals. Those who responded got a fanfare of nasal meows and demands for a good scratching. After Robin was laid to rest, however, Gregory began to decline.

He lost weight at a precipitous rate. His right ear was deeply infected and he wobbled around, falling in a heartrending way down the stairs he used to patrol with such verve. His kidneys began to malfunction and he developed tumors in his intestines. Only a few weeks after Robin's death, he himself grew too weak to continue his affection-seeking games. Reduced to skin and bones, he had to be coaxed to eat baby food from a spoon.

When Sybil arrived, Evelyn was just about to give the cat his daily dose of medicines, and she handed him over for Sybil to hold. Gregory's body felt as light and angular as a bundle of coat hangers. He crouched, uncomplaining, in Sybil's lap, leaning his head against her chest and waiting for the inevitability of the pill they were going to have to push down his throat.

Evelyn had a plump, young cat of her own. She loved Gregory, Sybil could see, but every time she had to touch him she guiltily tried to get it over with as fast as possible. After the medicine, Evelyn suggested that Greg might like an airing in the garden.

Behind Evelyn's house was a backyard, a surreal marriage of city and country. Under a canopy of feathery ailanthus trees, several faded canvas chairs sat in a tangle of flowers, as though in the midst of a bouquet. At one end of the garden, three fat goldfish fanned their tails in a sunken bathtub, and threading through it all, lending overtones of ancient Rome, was a white, mosaic pathway, which on closer inspection turned out to be made of broken pieces of bathroom tile.

The two women sat under the trees and Sybil held the cat. It seemed to her that his will to live was so weak that he could not possibly be around much longer. To comfort and communicate with him, she began doing small, slow Raccoon circles all over his emaciated body, and after a few moments he lifted his head, looked up at her, and then stretched his paws luxuriantly out in front of him.

"He seems to like that," Evelyn said. "I have to confess, I don't like touching him. It kind of scares me. I feel terrible about it, but that's the way it is."

Evelyn watched for a while, then asked to try the circles. She found that touching Gregory in a such a specific way seemed to completely alter her focus. The clear-cut and positive concentration demanded by the TTouch left no room for her instinctive impulse to recoil. Her fear evaporated, melting away in the face of the reality of connecting with Gregory.

A few days later, Gregory died. Evelyn had stopped the medication. "It always seemed forced anyway," she said. "He was weakening every day. Why trouble him?"

On his final day, Evelyn stayed home from work to be with him, and to hold and TTouch him when he seemed to want it. Stan, the longtime building superintendent, came to see Greg, and knowing it might be the last visit, Evelyn pulled the cat gently from his hiding place beneath the bed and held him cradled against her chest, heart to heart. At first he lay limp, then he gave a feeble purr and stretched. It was an odd, effortful, down-to-the-claws stretch, as if he was either grasping at or slipping out of something.

Stan had taken care of Gregory, too, in an unsentimental but very concerned way. Looking at him now, he shook his head and said, "Pets—people get 'em for their kids, so when they have babies they can show 'em the miracle of birth. What about the miracle of death?"

After he left, Evelyn laid Greg on a folded sheet in a box nearby so she could be with him while she did desk work. His breathing was shallow. He was nearly inert but something was still present. In the early evening, she glimpsed a sudden movement—he lifted his back leg as though to scratch his ear, something she thought he couldn't do anymore. He gave a kick, his head flipped back, and in that instant his remaining life seemed to fly out of his open mouth.

A few hours later, four neighbors, including Stan and Evelyn, came together as if on signal. Hushed and solemn, they went down to the garden in the dark, to the mulberry bush where other cats had been buried, and in the beam of a flashlight, they dug Gregory's resting place. As they lowered him into the hole, his ever unruly brown tail escaped the sheet in which they had wrapped him, and lay against the moist, cool earth like another kind of tree root.

The hum of fans and the clatter of dinner dishes being washed surrounded the group as they put in their handfuls of earth, said good-bye, and filled the hole. Then they stood quietly with their arms around each other, as if refueling the breath, pulse, and coursing of life through their bodies.

Later, talking about the experience, Evelyn told Sybil that Stan had been right—death was a miracle—and through Gregory she had been able to witness its phases without fear and with reverence.

11

Caring for Wildlife:
The Way Home

We will never find a way home until we find a way to look the caribou, the salmon, the lynx, and the white-throated sparrow in the face, without guile, with no plan of betrayal. We have to decide again, after a long hiatus, how we are going to behave. We have to decide again to be impeccable in our dealings with the elements of our natural history. . . .

What we do to animals troubles us . . . and our loss of contact with them leaves us mysteriously bereaved. If we could reestablish an atmosphere of respect in our relationships, simple awe for the complexities of animals' lives, I think we would feel revivified as a species. And we would know more, deeply more, about what we are fighting for when we raise our voices against tyranny of any sort.

—Barry Lopez
"Renegotiating the Contracts"
Parabola magazine

In the hill country between San Antonio and Austin, a free-lance trapper set out one morning for a day's work. Sometimes he was hired by the county, but on this particular day his employer was a rancher. His job is known as "predator control." This involves searching out the dens of coyotes, bobcats, and foxes, pouring gasoline down into them, torch-

217

ing them, and then pulling out the bodies of the mother and the young. The idea is to wipe out entire families to keep the predator population under control.

A few days after this particular "clearing action," a man who was taking a walk through the woods with his kids heard a low whimpering coming from beneath a stand of scrub cedars. When he went to investigate he found a little fox about the size of a three-month-old kitten, crying in front of a blackened hole, a lone survivor among the bodies of her brothers and sisters. Over 70 percent of the fur on the kit's body had been burned away, and the man tried his best to pick her up without causing her even more pain.

Both the little fox and the trapper were caught in what has become an escalating global problem—the collision between the needs of humans and the survival of the wilderness. Everywhere, as humans expand their domain, creatures of the wilderness find themselves displaced. Deer browse our lawns and eat our flowers, coyotes are spotted in parks and suburban orchards, the bodies of small animals lie scattered on our roadsides. Animals of the jungles and pampas are forced to vie with humans for life and land.

Fortunately, more and more individuals are becoming aware of this twentieth-century dilemma, and are demonstrating their concern. In proliferating animal rehabilitation centers and shelters around the world, people are learning how to care for the veritable Noah's ark of creatures who come to their doors injured, orphaned, or abandoned.

It was in one such place, the Wildlife Rescue and Rehabilitation Sanctuary, that I first met the little fox orphaned by fire. The man who found her had taken her home and nursed her. She was too wild to make an appropriate family pet, so he brought her to the sanctuary; by that time she had already bonded too closely to humans to be released back into nature.

Animals who are in this position are often troubled and difficult, but the little fox grew to hold a very special place at the sanctuary, becoming an invaluable go-between for the traumatized foxes brought in straight from the wild and the strange bipeds they were encountering for the first time.

Through the little fox's mediating presence, many of her brethren were able to relate to humans long enough to be helped and healed without having to abandon their wild natures.

"She was like a halfway house," sanctuary founder and director Lynn Cuny told me. "She even bred with an injured fox who had been hit by a car, and when the kits were old enough, we released them with the father.

"Where do you release the animals?" I asked, wondering why they wouldn't be caught or injured again.

"We are lucky enough to have access to a number of very large and protected private preserves," she said.

Lynn, a tall, athletic-looking woman in her late thirties, founded the sanctuary fourteen years ago when she quit her job with the San Antonio Zoo.

"I had become acutely aware that there was a large population of urban wildlife in the San Antonio area—raccoons, possums, skunks, foxes, and a diverse collection of birds. The trouble was nobody knew anything about them. When people found raccoons in their trash cans or skunks under their houses they would be horrified and call the police, who would come and shoot the animals. *Someone* has to do something about this, I thought, and it might as well be me."

The first years were lean ones—Lynn supported the sanctuary by driving a newspaper route from two to six A.M. Like many rehab centers around the country, she began hers at home and saturated the San Antonio area with her phone number, giving it out to pet stores, vets, zoos, fire and police departments, and county and state agencies.

Today the Wildlife Rescue and Rehabilitation Sanctuary is located on twenty-one secluded acres. Three thousand animals—mountain lions, jaguars, timber wolves, coyotes, deer, bobcats, foxes, raccoons, possums, squirrels, snakes, fish, birds, and retired zoo animals—come through the gates yearly to be aided by a staff of five employees and thirty volunteers. Because Lynn finds it practically impossible to turn any creature away, the staff can become overburdened, and the facilities crowded. She needs all the help she can get.

"The primary objective of the sanctuary is to get the animals that are viable released as quickly as possible," Lynn says. "Some, of course, like retired zoo animals or animals like mountain lions that have been declawed by their former owners, become permanent residents."

Lynn is deeply committed to the belief that animals marred by man "deserve our best efforts instead of little cages with concrete floors easy for us to clean."

The sanctuary enclosures are as different from cages as they can possibly be, constructed of fifty-foot-high telephone poles placed at regular intervals around an acre of land with fencing attached to create sides and a top. The result: slices of nature, with ceilings high enough to permit large oak trees and perimeters broad enough to include a varied terrain of dense brush, grasses, boulders, and even a stream.

As we walked around the land, Lynn introduced me to a coyote who had been shot in the hind leg. The wound had healed but he still exhibited symptoms of neurological damage. When upset, instead of running away he would spin around and around in circles. Lynn was interested in seeing if the TTouch could break this pattern, a symptom that would make it difficult to release him. The only way to catch him was to corner him and throw a blanket over him. Then he would lie totally still in the classic freeze and faint reflex, immobilized by fear (see page 111).

He lay at my feet with the blanket over his head, blanked out. As I knelt beside him, my intent was not to get into a relationship with the coyote himself—we wanted him to remain wild—but rather to deal with his injury on a cell-to-cell level. Though the fleshy part of the bullet wound had healed, the body's original response to the trauma had been to shut down neurological connections to the area.

I knew that if I were to begin working directly on the traumatized leg, the body would react by shutting down even more, so I went first to the corresponding area in the sound leg. Amazingly, a healthy area on one side of the body seems to resonate sympathetically with its counterpart—that is, if you affect one you also affect the other. The closest comparison is probably the example of two pianos side by side—play one and the strings of the other will sound in sympathetic vibration.

I began very gently to soften the hock and thigh area on the healthy leg using only a quarter of a circle. As soon as I felt the leg respond a bit, I switched to very light Raccoon and Lying Leopard TTouches. Then, having replaced the fear reaction with an acceptable connection, I was able to shift to the injured leg. Using the same progression of TTouches, I worked on the area, using the circles to remind the trau- matized cells that there is a more effective way to function. I finished the session by manipulating the entire leg in slow circles and flexing it back and forth. I slowly stood up to see what would happen next. When the blanket was removed, the coyote looked around for a moment, then jumped up and ran off in a perfectly normal straight line.

After I left the sanctuary, Lynn and the staff continued to work the TTouch on the coyote. He did well, and after a month was released into a fifty-thousand-acre preserve.

Some of the animals at the sanctuary are not as lucky and are left with permanent disabilities. Lynn asked me to work on a resident possum who had been hit by a car. She was paralyzed in the hindquarters and got around by dragging her useless back legs behind her. As a result, all the fur on the back legs had been rubbed away and the flesh was scarred and bruised. The poor thing was unattractive, determined, incredibly endearing, and still very much alive.

There was no way to cure this condition, we knew, but Lynn wanted to see if it was possible to use the TTouch to alleviate the possum's discomfort. Most of all, she wanted to communicate with her and let her know she mattered. For me, getting to know her was very moving. Here was a small animal most people would think of not only as damaged but worthless, being cared for in a wonderful natural habitat where she could hide behind rocks and under bushes and feel protected. I sat down with her, happy to feel beneath me not barren concrete, but soft, pine-needled earth.

Holding her in my lap, I used tiny Raccoon circles on her worn legs to make contact with her and synchronized our breathing rhythms in a technique called "pacing." At first I adjusted the flow of my breath pattern to hers, which was shallow and quick with anxiety. Once the connection was made and we were breathing in unison, I took the lead, quietly deepening and slowing my rhythm and feeling her follow into a calmer, much less fearful state, like slipping from cold rapids into a wide, still pool.

Gradually, though I had based our communication on an appreciation of our cellular sameness, I became aware of a very distinct individual, a being who was shy and sweet and full of an unusual degree of patient resignation.

With some species it's easy for us to lose sight of the individual nature of each creature; we tend to lump them all into one indistinguishable category, i.e. possum. Yet as I touched the animal lying so quietly in my lap, I experienced the absolute singularity of her life and the life of any being. This was a possumy self, expressing a heritage of round black eyes, rough, dusty-brown fur, and delicate paws, yet she was also matchless, as unique as a fingerprint, as individually alive as every daisy, every horse, and every human.

While Lynn's sanctuary in Texas is one of the largest in the United States, there are many others coast to coast doing the same kind of work. Near Rochester, New York, I went to visit Wildlife Rescue and Rehabilitation, Inc. to demonstrate how the TTouch can help calm animals and get them over the trauma of captivity until they can be released. A large number of raccoons arrive at their doorstep, and I worked with one of these roly-poly masked fellows using pencils and feathers as aids to the TTouch.

Pencils are to small animals what wands are to larger ones: a means to stroke creatures who are initially too frightened and aggressive for the direct contact of your hand. By using a pencil you can start your session with strokes that make a little more contact than those of a feather, with the additional plus that the animal can turn and bite the pencil, not you. I use the eraser end of an unsharpened pencil for making small circles from a distance. The eraser end is also good for making circles between the pads of small paws.

THE SAGA OF KENYON BEAR
AND THE MONTANA CUBS

When you handle an animal using the TTouch you are establishing a relationship of mutual trust rather than one of dependence. The animal keeps its integrity and doesn't become addictively fixated on you, and you in turn do not become dependent on the animal for exclusive love. Instead, the love that is exchanged is the kind with "no strings attached."

An important and controversial issue for many zoos and rehabilitation centers is how to resolve the problem of bonding: too much relationship and handling and you spoil the animal for zoo life or for release back into nature; too little and the animal suffers and is depressed. As I continue to work with people in zoos and rehab centers around the world, I'm finding that the TTouch seems to offer an effective solution to this prickly problem.

One of our most dramatic test cases came in the form of a five-month-old black bear cub orphaned by the great forest fires of Montana in the summer of 1988. His appearance in my life, along with that of twelve other little cubs, homeless refugees of the same firestorms, was sudden and poignant.

It was a desperate situation. The cubs, separated from their mothers and found wandering in the roads or alone in the brush, had become the charges of the Montana Department of Fish and Wildlife, but no money for their continued care could be spared from a budget already strained by record-breaking drought and fire. Unless an immediate plan for help could be devised, all thirteen cubs would have to be euthanized in a matter of days.

Hearing this news at a Fish and Wildlife Department meeting, a staff member immediately called Austin, Texas, to contact Eleanor McCulley, a TTEAM supporter and a member of Animal Ambassadors International, whom the staffer had met at one of my clinics.

AAI is an organization inspired by my first trip to the Soviet Union in 1984. While there, I began to dream of a grass roots foundation without religious or political affiliation.

Since then that dream has become a reality. With thirty-four member countries, Animal Ambassadors International is now actively engaged in programs that support a number of interrelated goals: the advancement of understanding and communication between species; the education of children in the importance of animals in our lives; the prevention of cruelty to animals; the demonstration of the therapeutic value of animals to humans; and the understanding of human-animal-environmental interdependence, to name only a few. Naturally, TTEAM is often an integral part of these programs.

So it was no wonder that Eleanor McCulley immediately thought of AAI when she heard of the plight of the Montana bears. The wildlife department had tried to find homes for the cubs, but they had run out of time. To add to their difficulties, one of the bears had sustained a concussion and neurological damage when he was hit by a car.

Eleanor called the department, and bought more time for the bears with the promise that AAI would take on the financial responsibility for them. There followed a flurry of phone calls, at the end of which Dr. William Sayre and his wife, Mary Anne, and Dr. William Rush and his wife, Joanne, of Austin, Texas, agreed to underwrite the project and raise the necessary money through donations.

Relief—the bears were saved. With this reprieve, the Montana wildlife people were now able to find homes for all but the little injured bear. He was weak and had muscle tremors in his legs that spread to his entire body when he was stressed.

So Kenyon, as we named him, was put into the care of AAI with the understanding that we work with him under the supervision of veterinarian Dr. Tom Beckett and his partner and TTEAM practitioner Marnie Reeder. Suddenly the frightened little bear became the focus of a team dedicated to his recovery, development, and eventual release back into the forest.

Dr. Beckett is the founder of the Camino Viejo Equine Veterinary Clinic in Austin, where the TTouch and TTEAM are used as a standard part of diagnostic, treatment, and care procedures. Tom and Marnie have written widely about TTEAM, interpreting it from a veterinary viewpoint, and applying it in innovative ways that have won them an award from the Humane Society of Austin and Travis County. Especially concerned about the problems facing wildlife rescue practitioners, they have introduced methods for using the TTouch on wildlife to rehabilitators around the country.

Kenyon was ensconced in the Rushes' big backyard in a fenced-in concrete area containing a large dog kennel, once the home of a Great Dane. He took one look around this alien world, ran to the kennel, and stayed there, too scared to come out. So we twined green cedar boughs in the chain link fencing, brought in logs and fresh foliage, and covered the cement flooring with soft straw. The flat roof of the doghouse looked like it might make a good bear roost, so we covered it with straw and piled bales of hay around it so that Kenyon could climb up on top if he wanted to.

The transformation of his environment worked wonders for the cub. He not only came out of hiding, he climbed up to his roof—undisturbed by the fact that a television crew had come to record his doings.

For the next three months Marnie worked Kenyon's body every day with the TTouch. At night Joanne Rush would listen for him. If she heard him whimpering she would come out to feed him or just sit with him and keep him company, sometimes even reading to him. Odd as it sounds, the rhythms of the human voice reading seem to soothe animals.

It was hard to resist the impulse to turn him into a pet, he was so cute and appealing. With his fuzzy head and paws poking out of his front door, he looked like an adorable illustration in a children's book. But baby exotic animals almost always grow up to assume their wild characteristics along with their grown-up teeth. Sadly caught between our world and their own, they become dangerous. Therefore it was both our goal and our challenge to heal Kenyon in an atmosphere that was warm and stressless, but that didn't promote bonding.

Baby bears need to stick close to their mothers for the first two years, and separation leaves them disoriented and prey to neurotic habits like pacing, obsessively licking their feet, and other compulsive behaviors. The foot licking is to replace the kneading motion they make when nursing, an action that simultaneously feeds them and serves to stimulate their paws.

To keep him from becoming fixed in these habits, Marnie worked the TTouch on his body and set up lightweight poles made of plastic plumbing pipes to interrupt the path of his pacing.

Marnie introduced these therapies slowly.

"Of course, when he first arrived he was not exactly a pushover," she says. "You couldn't just walk over and have at him.

"We started out by stroking him with the wand, making small circles with its knob end and then gradually moving our hands down the shaft until we could TTouch him with our fingers."

When he paced, Marnie broke up the behavior by placing the poles in such a way that he had to walk over them. At the same time, she gave him reassurance by touching him with a wand on either side of his body. Prior to the work, because of his neurological injury, one of his back legs was jerking uncontrollably out to the side and his front legs were not in alignment. After a few weeks of the poles and body work, though, he began to move with perfect efficiency and gained enormously in strength.

Early in Kenyon's Austin sojourn, I came to visit and to discuss the work with Marnie and Tom. Marnie had been avoiding his mouth and sharp little teeth, as well as his clawed paws, but if we were to proceed further with his rehabilitation, Kenyon would have to allow essential work on those areas. I wanted him to learn how to be careful with humans. He would have to learn fine motor control, how to differentiate between chomping down on a hand and gently holding it, between reaching out a paw with claws extended and doing the same thing with claws sheathed.

A bear's way of investigating or getting friendly is by testing or "tasting" with his or her mouth, and Kenyon had no way of understanding that his explorations could be painful to us. I began the mouth work by using two small two-foot lengths of hose to prevent him from biting. Every time he tried to bite I protected myself with the hoses, as well as overlapping my hands so that he couldn't get his mouth around them. After accustoming him to the hoses I began to stroke him gently under the chin, and down the neck and chest, using both hoses simultaneously in order to break down the pattern of resistance. Then, putting aside one of the hoses, I stroked him under the chin and along the sides of the jaw, with the hose on one side and the back of my hand on the other. When he readily accepted this touch, I moved on to using both my hands, loosely curling my fingers inward so that I was stroking the same areas with the midfinger knuckle joints.

Kenyon also felt threatened if anyone came near his ears, a vital area for TTouch work as we have seen, so I worked with one hand on his mouth and chin where he now felt comfortable and the other on his ears, and soon he accepted the contact.

When we started the session he was in his characteristic nervous behavior mode, keeping most of his body inside his house with just his head and forepaws sticking out, so I worked sitting in front of him on the ground. After the third fifteen-minute session, he suddenly reached up and touched my face gently all over with his warm, dry nose.

I held very still, and then, to my complete surprise, he gave me an amazingly sweet and refined demonstration of the trust that he now felt. First he touched each of my closed eyes and then with the delicate precision of a fine watchmaker, he took my eyelashes between his teeth and very softly pulled at them, first one eye, then the other. He did it so slowly, it was as though he was trying to reassure me and tell me not to be afraid, and it was one of the most moving moments of my life with animals. (See photo insert.)

Next, Kenyon and I addressed the question of how he could best use his paws when dealing with humans. To begin with, I kept one hand under Kenyon's chin, where he was now comfortable with contact, while stroking with the other hand down his legs and onto his paws. Bears put out their claws reflexively when reaching out to make contact, which is why when they are in captivity or working in circuses they are often declawed.

It really saddens me to see the way in which most circus bears are handled. Taken as young cubs, they are muzzled, declawed, and then taught tricks through a cruel system of physical punishment and reward. While it's important to respect and acknowledge danger when handling animals like bears, all of my experiences with animals over the years has shown me that violence isn't necessary and it's my great hope that we can evolve out of the notion that it is.

When bears reach out a paw in play or to examine something, the paw curves, which automatically extends the claws. Each time Kenyon reached out I would take his paw and hold it flat to show him how to keep the claws retracted and what that position felt like. I manipulated the paw to demonstrate for him the difference in sensation between claws retracted and claws extended. Then, holding his paw lightly in one hand, I worked the TTouch circles directly on and in between the pads. After a while, he put his paw very gently, claws retracted, onto my hand.

As with the mouth work, after several sessions he had understood what I was asking and what humans need in order to be able to communicate with him.

After three months in the Rushes' yard, Kenyon's adorable baby-bear charm began to turn into the scrappy antics of a healthy adolescent. Strong and well now, he was outgrowing his dog kennel headquarters and would tear around the place like a fine young horse feeling its oats. It was obvious that the time had come to return him to the wild.

We were delighted to see that while he had thrived on our contact with him, he had not bonded with any of us and was therefore ready for the next phase, a kind of halfway house that would simulate his life in the wild and allow him to learn to fend for himself.

Many people think that all you have to do is release an animal back into the woods and it will revert to its wild ways. Not true. When humans feed and shelter animals, the creatures forget their foraging skills and need time to relearn before they can be turned loose.

At the Wildlife Rescue and Rehabilitation Sanctuary, near San Antonio, Marnie oversaw the building of a large enclosure strong enough to hold the now nine-month-old cub. Kenyon saw no humans for three months. Food was hidden for him while he slept and an artificial pool was stocked with fish. After that period of isolation he was strong and self-sufficient, a wild bear again.

He was flown back to Montana where the wildlife department had thought they would release him. Unfortunately, it was not possible to find a suitable place for him, and at the time of this writing we are still searching for an appropriate home for Kenyon. The danger is that time is against him; the Department of Wildlife doesn't have the facilities to keep him in a wild and independent environment, and with every passing day in confinement he loses more of his ability to return to freedom.

As Kenyon's case demonstrates, the problems of rehabilitation do not always end when an animal is ready to be reintroduced into its natural habitat. In some cases, appropriately placing an animal that can be potentially dangerous brings on a whole new set of legal and other obstacles. Many park and preserve administrators believe that hand-raised animals pose a risk to humans. Such animals are friendly and unafraid of people, but when they ask for food and are denied it, they can become aggressive.

The predicament in our parks and wild areas is rooted not only in the behavior of bears but in that of humans. Because we lack knowledge, we create dependencies that later backfire. We feed them even when signs tell us not to. We encourage them to beg beside the road so that we can record their cute ways with our cameras. They seem so tame standing there, like giant living teddies—I even heard of a man who got out of his car and tried to put his baby on a bear's back to take a souvenir picture. It's when they don't behave like teddies that we suddenly, and sometimes horrifically, see firsthand the problem we've created.

In areas where expanding human populations move into bear habitat, careless garbage disposal brings the bears into a harmful and potentially dangerous dependent relationship with us. In areas like New Mexico, where encroaching humans and recurring drought have combined to drive the bears down from the mountains, rangers relocate them, moving them farther away to areas that then become overcrowded. The solution to these dilemmas is complicated, but as we begin to understand more about our own crucial role in the equation, surely answers will begin to unfold. It's up to us to make the effort to find them.

AFRICA'S ANIMAL ORPHANS

Shortly after the rescue of "the Montana thirteen," I got a phone call from Carolyn Bocian, then a keeper in the primate department of the National Zoological Gardens in Washington. We had talked often of her dream of starting a retirement center for primates released from testing laboratories. Now here she was, phoning to tell me of a chimpanzee rescue operation in Zambia that desperately needed help.

David and Sheila Siddle, a British couple in their mid-sixties, have lived in Africa for forty years. Sheila arrived as a young girl with her parents and David began his African sojourn as an engineer, later leaving that life to become a cattle rancher in Zambia. Over the years, they watched with growing alarm as Africa's wildlife became increasingly beleaguered and the traffic in baby chimpanzees became a thriving illegal trade across the continent.

Poachers, many of them poverty stricken as well as angered and confused by new conservation laws forbidding their ancient hunting rights, were shooting or poisoning adult chimps in order to sell the orphaned babies to the lucrative international pet and biomedical research market.

In the struggle against the poachers, the African governments retrieved many of these babies. Returning the young chimps to the wild, however, was like sentencing them to almost certain death or recapture, and so in 1983 the Siddles founded the Chimfunshi Wildlife Orphanage, one of the few sanctuaries for chimpanzees in Africa that could provide a refuge for these little orphans.

Without the Siddles' haven, the government retrieval programs had nowhere to send the orphaned chimps yet, Carolyn Bocian told me, David and Sheila were having problems maintaining their sanctuary. They badly needed funds both to continue their operation and to finish constructing a seven-acre outdoor compound where the orphans could live as a normal social group.

For three days after my conversation with Carolyn, the Siddles and their chimps stayed on my mind. I was actually writing them a letter requesting more information and discussing a possible adopt-a-chimp program for our TTEAM newsletter when another phone call came. It was Harriet Crosby, founder and director of the Institute of Soviet/American Relations.

"The institute is running so well," she said, "I actually have some time on my hands. So I thought if you're planning to travel somewhere to do something with animals, I'd love to come along."

Here it was, a possible answer for the Siddles. If we went to see them perhaps Harriet would be able to find grant money for their cause through her foundation contacts.

"How would you like to go to Zambia with me and check out an orphanage for seventeen chimpanzees?" I asked.

And Harriet, clear-eyed, straightforward adventurer that she is, was practically packing her bags before we had hung up the phone.

Getting to the Siddles' remote ranch in northern Zambia proved to be almost as hard as finding the headwaters of the Nile, but we finally made it. David and Sheila were relieved to see us. They had known we were en route but had lost contact with us because there were no phones and the telexes were down for the Christmas holidays.

My journal for our first day reads:

> We had a cold beer sitting under a long thatched roof with handmade table and chairs. The view is spectacular over a river flood plain with several hundred acres of grass bordered by forests as far as the eye can see. In the rainy season the river overflows and the plain is covered with water. A lawn stretches out from the sitting area. Left of the lawn is an enclosure with twelve orphaned baboons waiting hopefully for release. An orange orchard borders the far side of the lawn where geese and wild ducks strut.

On the right is the house and all seventeen chimpanzees. We meet them and are warned to keep the Tusker beer glasses out of reach as they love to grab them.

The nine youngest chimps are on the outside of the house, the older ones are inside in a long, wire mesh enclosure that actually forms one entire wall of the living room. They meet us gently, reaching through the mesh to take our hands in theirs. Coco snakes out nimble fingers and grabs my money out of my pocket. Sheila rescues it just in time. Sandy, a four-year-old, has a cold. I work his ears and he leans against the wire and sticks his ear out to me.

When a chimp is sick, if it's young enough to diaper, it is taken into bed with Sheila and Dave. If not, Sheila sleeps in the straw with it.

5:00 PM:

Patrick, the caring Zambian keeper who has been with the chimps for five years, goes off duty. Sheila and David feed everyone their ball of cooked cornmeal mixed with vegetables and garlic. All the chimps are very gentle and orderly in receiving the food.

5 to 6 PM every evening is escape hour when the chimps try to break out, so David must go carefully around all the cages checking the wire. We make a tour of the seven-acre enclosure. Only the gate, the holding areas, and the electrified wire remain to be finished. The area is luxuriously wooded and grassed. An enormous wall surrounds it, twelve feet tall with electric wire planned for the top as much to keep the poachers out as to keep the chimps in. There will be a lower strand of electric wire at five feet. David has put it there because he says the chimps will most likely break off branches and lean them against the wall to scale it. He wants them to encounter the first electrified wire before they get to the top one where they could fall such a long way back down and injure themselves.

We return and sit at the outside table in the twilight. Sheila is interested in the TTouch not only for the animals but for humans. She treats all the Zambians in the ranch compound, a village of 250 men, women, and children who are the Siddles' personal responsibility. A nine/tenths full moon lights the African night. We go inside for dinner and afterward get ready for bed. The generator has been turned off for the night and we wash up in the peaceful glow of kerosene lanterns.

The alarm is set for 6 AM so that we can accompany the younger chimps on their exercise trip with Patrick. Each day they venture into the forest for seven hours to walk, run, play, sit in the trees, eat fruit, and just relax. Sheila suggests we sleep in and follow them later but we've come halfway around the world to find this orphanage. As Moshe [Feldenkrais] used to say, "I can sleep when I'm dead."

In the morning Harriet and I took off into the forest with Patrick, the nine chimps, and Mark and Tracy, two young British volunteers who had only recently arrived at Chimfunshi. Mark had worked with chimps before as a volunteer zookeeper, but Tracy's experience so far had mainly been with sheepherding, sheepdogs, and horses. Right from the very start she was a bit nervous, which is probably what set the stage for the trouble that was to come.

With some of the chimps clinging around our waists and others cavorting around us we paraded off down the trail. Sunlight filtered down through the tall trees that stood widely spaced among high, lush meadow grass. Sandy, Tara, Tober, Jane, Donna, Rita, Cora, Boo Boo, and Coco, the nine youngsters, spread out on both sides of the path. They ran to the giant orange mushrooms that grew everywhere and slurped up the rainwater that had collected in their concave tops. They zipped up trees to bring down nuts and fruit, and played tag, running up and down the twenty-foot-high anthills that were everywhere, giant ant cities under earthen domes. After thirty minutes we stopped in a glade to rest.

That is to say, we *thought* we were going to rest, but the chimps continued the fun, raining down on our heads from the branches above, making swinging nests in the treetops, and inviting us to tickling matches. Harriet and I, totally captivated and infected by the spirit of hilarity, were coerced into their favorite game, swinging enthusiastic chimps around by their arms and legs.

Then, with the suddenness of a tropical storm, the mood changed. Tracy, the British visitor, and I had sat down on a log when jokester Donna raced by and made a grab to steal our backpack. Tracy yelled and made a lunge for the pack, which upset Tober. Thinking Donna was under attack, he came dashing over protectively and bit Tracy on the calf.

Tracy's face turned white with shock as two blue holes appeared in her leg. Her calf went into muscular spasm. She was in considerable pain, but there was no external bleeding. While Harriet worked Tracy's ears for shock, I went to work on the normal leg to release some of the fear (see page 140). Then moving to the injured area, I encircled the wound at a distance of six or seven inches with Lying Leopard circles, after which I came in closer to within an inch of the wound and switched to the Raccoon TTouch. Within fifteen minutes the pain was reduced by at least half and the wound was shrinking before our eyes.

Thirty minutes later the tooth marks were covered over with light scabbing and had faded to one-quarter of their original size. The pain had disappeared completely.

The procedure wasn't unfamiliar to me. When you work closely with animals, chances are that once in a while one of them will catch you with a bite. Should you get bitten, the best thing to do first is to clean out the wound with an antiseptic, and then apply the TTouch, never directly on the puncture but always around it as I did with Tracy. With bites there is always a danger of infection, which is why it's so important to clean out the wound as soon as possible. Sometimes, however, the incident happens when you have no first aid available. To our astonishment, we've discovered that nine times out of ten when the TTouch is applied immediately, there is no ensuing swelling or infection. But once again, it is preferable to use antiseptic first, if it is available.

When we went home from the forest that night, the bite on Tracy's leg was well on its way to healing, but her psychological wound was not. Patrick was angry with Tober and Tracy was tense and nervous around the chimps.

Tober, a two-year-old, had been captured as an infant and sold to an American family who had hand-raised him. When they realized that he was getting too big for them to control they gave him to the Siddles. He had more to deal with than the other chimps because he was trying to adjust not only to sharing human love and attention but to being a member of a chimpanzee social group as well.

Another of the Siddles' young chimpanzees, Charlie, had similar problems. Always beautifully behaved with the Siddles and Patrick, he reacted badly to visitors who didn't understand chimp behavior. As a result, Charlie was now kept caged. Harriet and I were worried that the same fate could be in store for Tober. We decided to try to get Tober and Tracy to work together.

The opportunity to do so came about dramatically and almost immediately. Two days later, Tracy and I had spent the morning training a three-year-old mare, so when we arrived in the forest the chimps were already there, browsing and playing. Once again it was Donna who precipitated the action.

She came running to greet us with Tober right behind her. Tracy was still nervous from two days before, and seeing the chimps come flying straight for her, she unconsciously reared back and stiffened her posture in fear. In chimp communication, such erect, stiffened posture is a message of anger and threat, and Tober once again thought that Tracy meant to harm Donna. In a flash he leapt forward and bit Tracy in the thigh.

This time it was too much for Patrick. He grabbed a stick and chased Tober with it, yelling at the chimp in a stream of furious Zambian. Tober took off and as he ran, he turned around once like a kid full of frightened bravado, to wave his arms and bare his teeth at Patrick. This made Patrick so mad he threw the stick and hit Tober hard in the side, sending the chimp screaming in protest and terror up the nearest tree.

Patrick's anger was a terrible thing for Tober. All the chimps had a high respect for the tall, imposing Zambian and regarded him as pack leader, Big Daddy, and head honcho all rolled into one.

Once again, Harriet and I immediately began working Tracy's ears and wound. After thirty minutes she had calmed down and Tober crept out of the tree. He was miserable, hanging around nearby like a child who has been bad and wants only to be forgiven and understood, but Patrick wouldn't let him near us. Finally Harriet and I convinced him that it was important for us all to let the chimp approach.

But Tober was too confused by what had happened. He watched us with worried eyes, torn between his fear of us and his desire to be back in our good graces. It took fifteen minutes to coax him over. In my experience with primates, getting down lower than they are and turning your back on them is a way of communicating your submissive and peaceful intentions, as is grooming or eating in their presence. In order to ease Tober's mind, I got down low on my hands and knees and crept around, frequently turning my back on him, while near us Harriet lay in the grass and groomed Sandy. As with many animals, when primates feel frightened, direct eye contact is often interpreted as a threat, so I kept my eyes down. When I did glance at the chimp, I made my gaze soft and unfocused.

Finally Tober slipped shyly over to me and let me groom him with circles and gentle fur pulls. The next step was to coax both Tracy and Tober into sitting together. That accomplished, we got Tracy to groom Tober using the Lying Leopard and Lick of the Cow's Tongue (see pages 242 and 250), two TTouches that are very soothing both to giver and receiver.

Gradually, the fear between them melted. Tober climbed up into Tracy's lap and put his arms around her neck, wanting her to know he hadn't meant to harm her. By the end of the day Tracy had become Tober's special friend, and it was to her protective arms that he ran when a snake scared him, and her hand that he chose to hold on the trail going home.

Having seen the Siddles' situation with her own eyes, Harriet was able to arrange a foundation grant not only to complete the enclosure for the chimps but to finish a large new kitchen for Sheila, who had been sweating it out cooking for seventeen chimpanzees and ten people in a twelve-by-twelve-foot space.

But what will happen to the Chimfunshi sanctuary when the Siddles are no longer there? How can their dream of a two-thousand-acre preserve ever be realized in an Africa where animals are the focus of so much violence and confusion, of so many conflicting traditions and motives? How can a poacher who gets the equivalent of a year's salary for one animal understand why the capture or slaughter of an elephant, chimpanzee, or gorilla is a tragedy? How can a starving people comprehend the fact that conservation is not a matter of restrictive laws and faceless bureaucrats, but a vital question of their own survival and the survival of their children?

For the last decades we have been moving toward ecological disaster, and there are many voices that say it is already too late to turn back the tide we have loosed upon ourselves and upon our fellow inhabitants of what Buckminster Fuller used to call "the spaceship earth."

In the face of such an uncertain future I have had to ask myself if anything is really accomplished by spending so much time, money, and effort in doing things like rescuing a small band of chimps in Africa or a single baby bear in Montana. The complexity and magnitude of the changes needed sometimes overwhelm me, as it probably overwhelms many of us. Then the thought comes that like drops in the ocean, each person, family, and group is a reflection of the whole of our planetary life, and when we reach out as individuals to make a difference, we are each one, reflecting what is possible for that whole.

I believe that when we open our hearts to animals and care for them in whatever way we can, we create a ripple that spreads from our single action to affect everything else on the planet, the whole inextricable, living web of plants and stones and creatures and humans that we call home.

And I believe each one of us can make a difference.

—— 12 ——

How-To:
A Practical Illustrated Guide
to Techniques and Terms

TTEAM: TEAM was originally the acronym for Tellington-Jones Equine Awareness Method, the name given to the system of training and body work for horses and riders that I spent so many years developing. After my Feldenkrais studies, as I continued to explore the specific possibilities of equine body work, the TTouch emerged. We found that while the TTouch could be used as an independent system, it also remained an integral part of the training methods we had begun to call TEAM-work. To indicate that TEAM employed TTouch we added an extra T, making TTEAM.

Eventually, as we began to adapt the TTEAM methods to work with dogs and cats and cows and bears and myriad other species, we took on an additional definition for our acronym: the Tellington-Jones Every Animal Method. And of course the name represents the goal of our work, to help you and your animal truly become a team, with all the communication, fun, and cooperation that the word implies.

TTEAM is currently being practiced in over thirty-four countries.

TTouch: We use the double T not only because it stands for Tellington Touch but because the abbreviation TT is also the Greek letter *pi*, which is specifically related to circles. For the basic TTouch, rest the thumb and little finger on the animal as support, and, using the middle three fingers, with the center finger leading, start at number six (the bottom of the circle on your imaginary clock) and push the skin around in a circle and a quarter—then release. The hand and arm should remain flexible, and the practitioner should be aware of his or her breathing (see *Refining the Circles*, page 238).

When working on small animals, for example a small bird, a little lizard, or a young kitten, you may choose to use any one of the three fingers rather than all of them together, depending on how your hand is positioned. To work between a bird's feathers or on a kitten's ear, for instance, you would use just the tip of one finger.

We start the circle at six o'clock in order to initiate it with a lifting of the skin; if you begin at twelve o'clock the fingers pull down and tighten the skin rather than releasing it. The circles are intended to awaken the function of the cells and activate neural impulses, thereby bringing a new awareness to the body.

Many people mistakenly think of the TTouch as "massage" because it is a word they can relate to, and that is what it appears to be until experienced personally. Most massage techniques involve a rubbing motion and are intended to affect the muscular system. The TTouch *moves* the skin rather than rubbing it. The intention is to open unused neural pathways to the brain and to activate the function of the cells. Of course, when you affect the nervous system you are also affecting muscle.

REFINING THE CIRCLES

I would like to elaborate here on the description of the basic circular TTouch in Chapter Two (page 23). When my sister Robyn was teaching the circles she noticed that it was very common for beginners to stiffen their hands and fingers when practicing the TTouch circles. Try stiffening your fingers and making a circle and you'll see that it tightens your diaphragm and constrains breathing.

When Robyn tried to explain how to hold and move the fingers in a more effective way, she had a hard time finding the right words, until one day while looking through an anatomy book she came up with a great little rhyme scheme.

Your fingers are divided by three joints into sections called phalanges. The phalange from the tip of your finger to the first joint is called the distal intermedial phalange. The section from the first joint to the middle joint is called the intermediate intermedial phalange, and the section from the middle joint to the knuckle is called the proximal intermedial phalange. Why not call the three phalanges DIP, MIP, and PIP, Robyn thought. The names sounded like characters in a cartoon and would be easy to learn and remember.

When most people first try the TTouch, the tendency is to keep the DIP, MIP, and PIP joints straight and stiff and then attempt to push the skin around in a circle without any movement in the fingers. Try it with a very light pressure; the fingers have a tendency to slide across the skin. Now try the same circle with the DIP, MIP, and PIP joints rounded and flexible and see how softly and easily you can push the skin around when all three joints move with the motion. Try making a bigger circle; you'll see that although your fingers remain flexible with the motion, you'll need a slightly firmer pressure to keep from sliding. So the size of the circle you make depends on the looseness of the skin.

To make the circle itself, imagine the face of a clock as described on page 24. Position your thumb and little finger as anchors resting softly against the body of the recipient. Then place the other three fingers lightly touching each other at six (the bottom of the imaginary clock) and begin your circle with the middle finger leading the movement. Most people have a tendency to begin the circle quickly, rushing the fingers from six to nine and then slowing down. Instead, take your time through seven and eight, keeping the circle round, the pressure and pace even. In the beginning of your practice it helps to visualize each number of the clock as you approach it. If the skin is tight on an area of the animal you may have to slip very slightly over a part of the circle in order to keep the roundness, whereas if the skin is loose it's easy to push in a circle without slippage.

As you move from six toward seven, the first DIP joint reaches out and continues reaching out so that by the time you reach twelve the DIP phalange is lying flat. It remains flat as you move toward three and then raises again past three as the DIP and MIP joints flex to pull the skin around back to six. As you move past six to eight you reach out again with DIP, then at eight you pause and release.

Start your practice by trying a few circles on your own hand or forearm. Make a circle and a quarter in one spot and then move on to another spot at random and repeat. As a rule we suggest that the circles be done clockwise, because as acupuncture techniques verify, the clockwise motion is a strengthening one while the counterclockwise direction disperses tension.

The direction in which you make the circles usually remains the same regardless of whether you are left- or righthanded. Sometimes, when a person or animal can't stand to be touched, has an injury, or is in pain, the counterclockwise circle becomes acceptable whereas the clockwise one does not, making it necessary to reverse the usual procedure.

There are times, too, where you'll find yourself instinctively changing over from one direction to another. Trust yourself when this happens and just continue on—your intuition is usually right.

Although we have been working with TTouch circles and the direction of these circles for many years now and with many hundreds of people, I still don't have a final answer for when to use clockwise and when counterclockwise, because I find there are always exceptions.

TTouch pressure: The scale of pressure from one to ten in which the circles are executed. To determine the degree of pressure represented by each number, do the following exercise: place your thumb lightly on your cheek and touch your eyelid with your index and middle finger; make the contact as light as possible, just enough to move the skin over the eyeball in a circle. Transfer your fingers to your upper forearm and move the area with the same amount of pressure as on the eyelid. Make circles with the pads of your fingers in several places on your arm to get the feel of the pressure. You should notice there is no indentation on the skin. That is a one pressure.

Go back to your eyelid and make several more circles with as much pressure as is *comfortable* (do this on your forehead if you're wearing contact lenses). Repeat the process on your forearm and notice that this time there is a slight indentation in the skin. That is a three or four pressure. Pressing three times as deeply as that will give you a nine or ten. If you believe that this is not a great deal of pressure, remember—our intention is to affect the nervous system and the cells rather than the muscles themselves.

FIFTEEN TELLINGTON TTOUCHES: WHEN AND HOW

We have given the various Tellington TTouches and leading positions names inspired by the animal kingdom because it helps to make them easier to visualize and therefore easier to learn and remember.

The following TTouches employ the single clockwise circle (except when used counterclockwise, for situations described on page 44).

THE CLOUDED LEOPARD

When: The basic Tellington TTouch is the Clouded Leopard, named after a clouded leopard I worked on in the Los Angeles Zoo. The "cloud" part of the name describes the lightness with which the whole hand contacts the body, and the "leopard" stands for the range of pressure of the fingers. A leopard can be very light on his feet as in the light TTouch of a one, two, or three, or very strong, as in the eight to ten pressure scale. The stronger Leopard TTouch is appropriate for the more heavily muscled or blocked animal.

How: Hold the hand gently curved. Using the pads of the fingers, do the circles as described above and on page 25

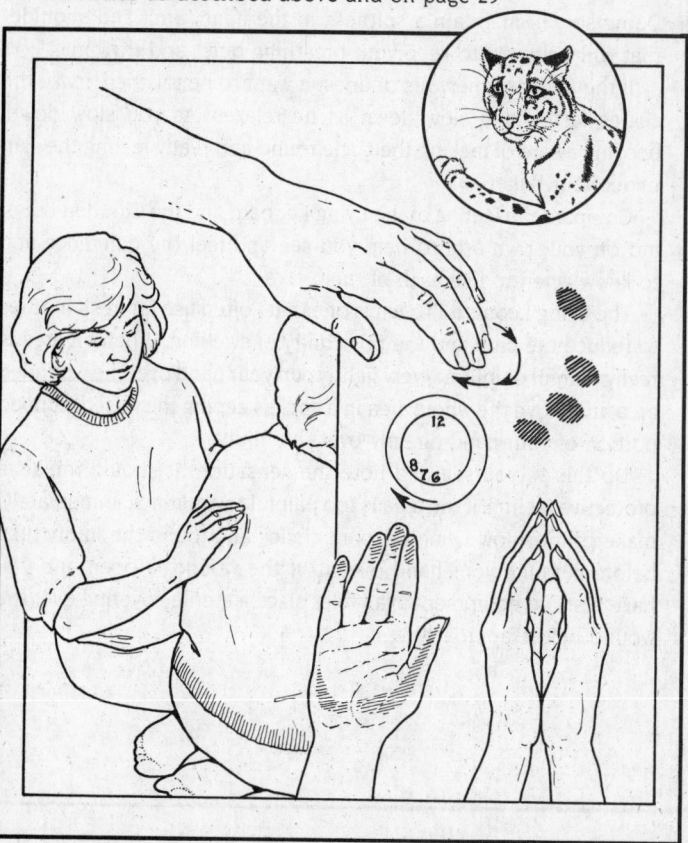

THE LYING LEOPARD

When: This TTouch is good for sensitive animals when the Clouded Leopard is considered too invasive or threatening. Good, too, to reduce the pain and possible swelling of injuries.

How: For this TTouch the leopard lies down, i.e., the curve of the hand flattens somewhat, allowing a larger area of warm contact. Be sure to allow your MIP and PIP (see page 238) knuckle joints to move, following the fingers (don't move the DIP joint, however). Doing so will maintain a softness in the hand, arm, and shoulder that will help you to keep your breathing quiet and rhythmic.

If the animal is nervous or doesn't want to be touched, make the circles faster, and slow down as he relaxes. As you slow down, become aware of making the circle round and really feeling the skin under your fingers.

Compare the feeling of the Lying Leopard and the Clouded Leopard on your own arm to help you see and feel the difference and to know when to use each of them.

The Lying Leopard TTouch is one that I often use for fresh injuries to reduce the pain and the possibility of swelling. When an area is really painful or injured, very lightly cup your hand over the wounded area and move the whole area in a circle, keeping the raised, cupped portion of your hand directly over the injury.

Do this to yourself and note the sensation: it should impart a protective feeling. If the area is too painful to approach immediately, make gentle, slow Lying Leopard circles all around the injury first, before cupping your hand over it. If the wound is open and you have first-aid equipment available, place a sterile covering over the wound before approaching it.

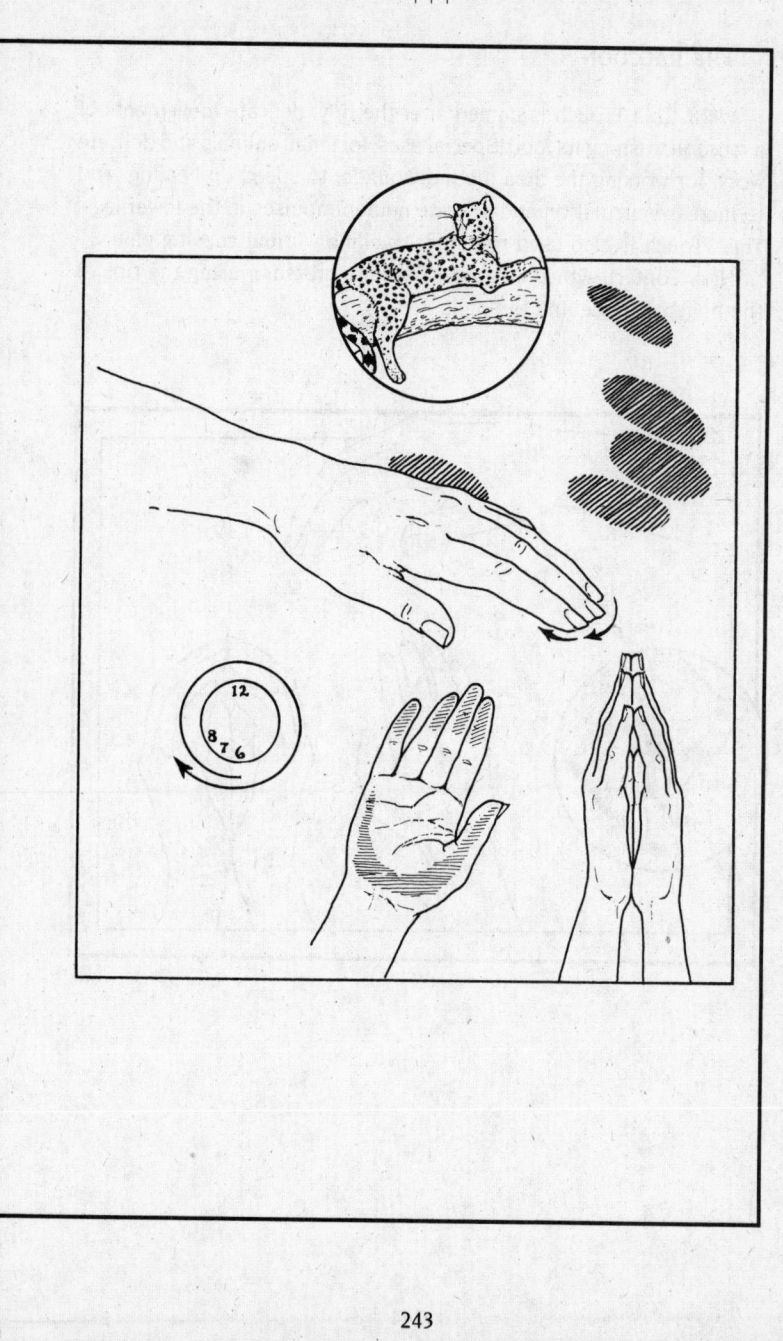

THE RACCOON

When: This TTouch is named after the tiny, delicate movements of a raccoon washing its food. Special uses: for small animals and delicate work; for working the area around wounds; to speed up healing; and to increase circulation and activate neural impulses in the lower legs. This TTouch is also used to reduce swelling without causing pain.

How: Contact with the lightest possible pressure, using the tips of the fingers just behind the nails.

THE SNAIL'S PACE

When: Watching a snail cross my path one morning gave me the name for this TTouch. The contractions and extensions of her body as she made her way reminded me of the small lifting and releasing motions of the fingertips we use to relax back and neck muscles, improve breathing, and reduce stress. *For horses*: when used on the back and the top of the neck it helps in lowering the head.

How: Using the fingertips to the DIP joint (see page 238), gently move the muscle upward half an inch to one inch, hold a few seconds, and then *slowly* release back down again.

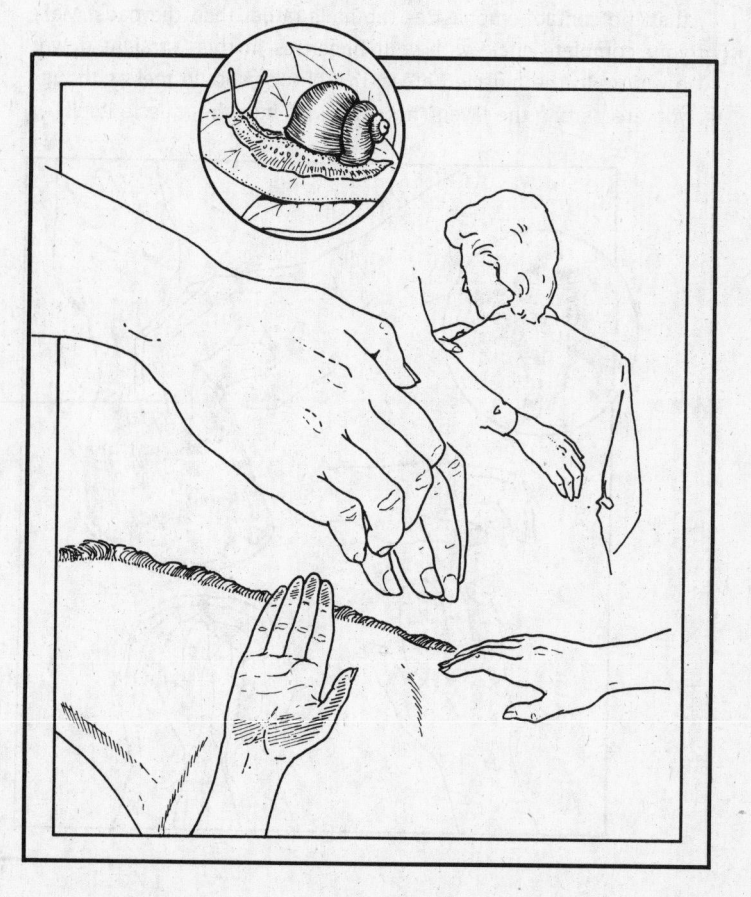

THE BEAR

When: The Bear TTouch allows the fingers to move deeply into areas of heavy muscling without discomfort to the recipient or the practitioner.

How: Your fingernails should be medium length, somewhere between one-eighth and one-quarter of an inch long, so that when you direct your fingertips straight down the recipient can feel the nails. Place your fingers on your imaginary circle at six o'clock so that the contact emphasizes the nails rather than the pads. Make your complete circle with your fingertips in this "straight down" position. In the muscled areas, the TTouch should feel as though you are parting the layers, not "digging" into the muscle itself.

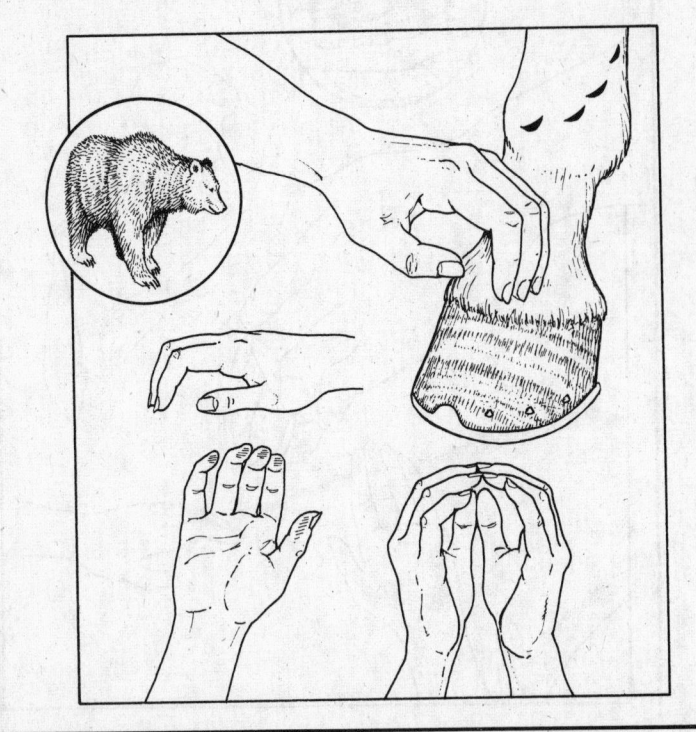

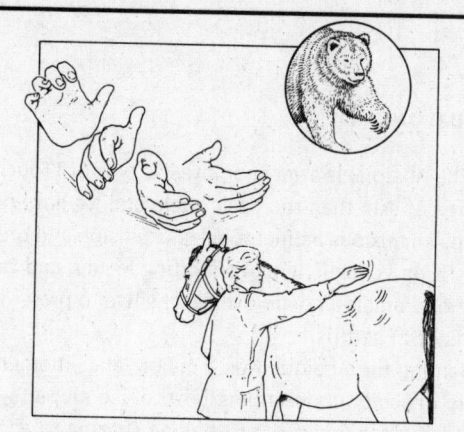

A *Refinement of the Bear*: **THE FLICK OF THE BEAR'S PAW**

When: A useful TTouch for approaching animals with initial fear of contact. We also use it for performance dogs and horses who have been worked with the TTouch just prior to entering the show ring or arena. After the calming effect of a TTouch session, the Flick of the Bear's Paw wakes up the body and makes the animal alert for the event without creating nervousness. If you don't finish the session in this way the animal can be too relaxed to perform well.

How: This TTouch is a cupping stroke that is used randomly over the body. The motion reminds me of a bear fishing for salmon or a person brushing lint off clothing. The amount of pressure used in this TTouch will vary from animal to animal. Some animals like a vigorous movement and contact while others prefer a very light almost sweeping touch. Use your wrists flexibly as you skip from place to place on the body. The animal may move around a bit at first but generally soon becomes quiet.

FEATHERING

When: A lighter version of the Flick of the Bear's Paw, used when the animal is very fearful of being touched.

How: Brush as though you had feathers on the ends of your fingers, so fast that your fingertips are gone almost before they arrive.

THE ABALONE

When: The Abalone is a very noninvasive type of TTouch. It moves a larger area of skin than the basic circle and we have often found that it helps an anxious animal to release tension and breathe more slowly. Try it on yourself. It's a comforting feeling and can be used on people or animals who are overly sensitive to pressure from the finger pads or fingertips.

How: Naming the Abalone was a major "aha" moment for me. I always had difficulty in describing how to use all parts of the flattened hand, but the image of an abalone sticking to an ocean rock was perfect.

Visualize your whole hand sticking to an animal's skin, like the abalone, moving the mass of skin or muscle in a soft circle, with just enough pressure so that your hand does not slide over the surface of the skin but actually moves it. Start the circle with your fingertips at number six on your imaginary clock face and move around the face of the clock to nine, twelve, three, past six, to eight. When you reach eight, instead of releasing, go back to six again and then gently lift your hand. If you are practicing on another human, and he or she is wearing clothing made of a slippery material, wet your hand to prevent it from sliding.

Note for horses: In the case of pain caused by acute navicular or laminitis you may only be able to make a quarter of a circle (six to nine and back to six again).

Note for long-haired animals: The Fur Pull: Slide your fingers beneath the fur so that you are holding it between your fingers, and then lightly and softly move the fur as you move the hand. Be sensitive to how hard the animal likes the pull. We've found that almost all animals really love this "hair pull," the human animal included.

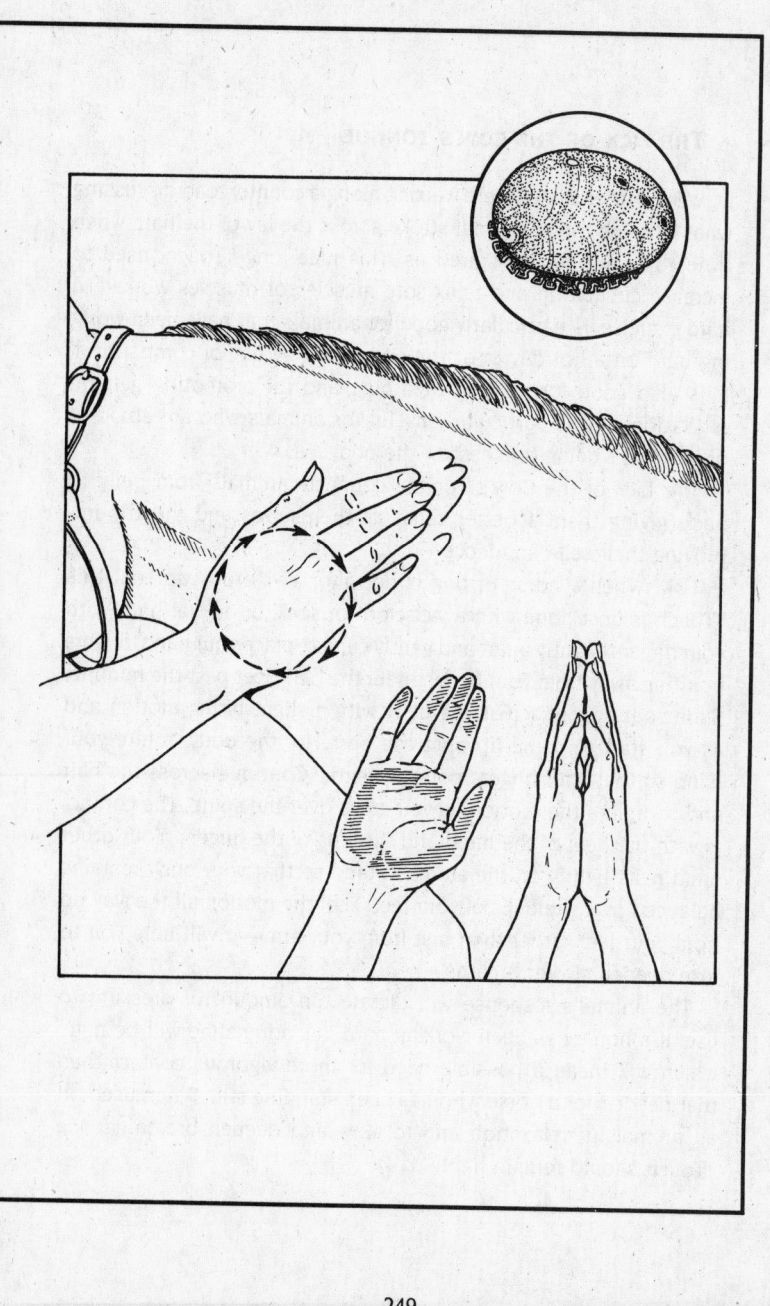

THE LICK OF THE COW'S TONGUE

When: A friend and I were walking along a country road discussing what to call a long diagonal stroke across the lay of the hair, when a nearby cow's moo inspired us. This nice long TTouch, used to increase circulation and relax sore muscles or muscles worked in hard exercise, is particularly good for animals that have been working hard on a hot day or at the end of a long day of competition. It is also good for slowing breathing and for promoting general relaxation, and is particularly useful for animals who are afraid of or don't like being touched on the body.

The Lick of the Cow's Tongue connects animals from belly to back, giving them a better sense of themselves and thereby improving their self-confidence.

How: When a horse or dog is hot after a vigorous workout this TTouch is best done with a wet cloth or sock, or on wet hair. With your fingers slightly apart and gently curved, place your hand, fingers pointing away from your body, under the belly just past the midline. Slide your fingers across the belly with a slight lifting motion and as you start to come up onto the barrel of the body, rotate your hand so that your fingers point upwards. Continue across the hair and complete the motion as you cross over the spine. The contact is with the heel of the hand and the tips of the fingers. Your other hand rests lightly on the animal. Stand so that your body remains balanced over the balls of your feet. Feel the motion all the way up from your feet rather than just from your arm—it will help you to take deeper, slower breaths.

The animal's response will dictate the amount of pressure to use. If a horse has been working hard, his adrenaline will be high, which will mean a possible need for more vigorous contact than that needed for a horse who has been standing still. When used on an animal for relaxation and to slow and deepen breathing, the TTouch should remain light.

For horses or other large animals, this movement can also be executed across the shoulder, starting at the lowest point of the shoulder and moving up to the withers (the top of the shoulders).

Sometimes, when working with an overly sensitive animal, the skin will twitch. When this happens, stop your movement and do a light Abalone circle before moving on the next few inches to the next area. This encourages the animal's calm breathing and breaks the habit of contracting the muscles and pulling away from contact.

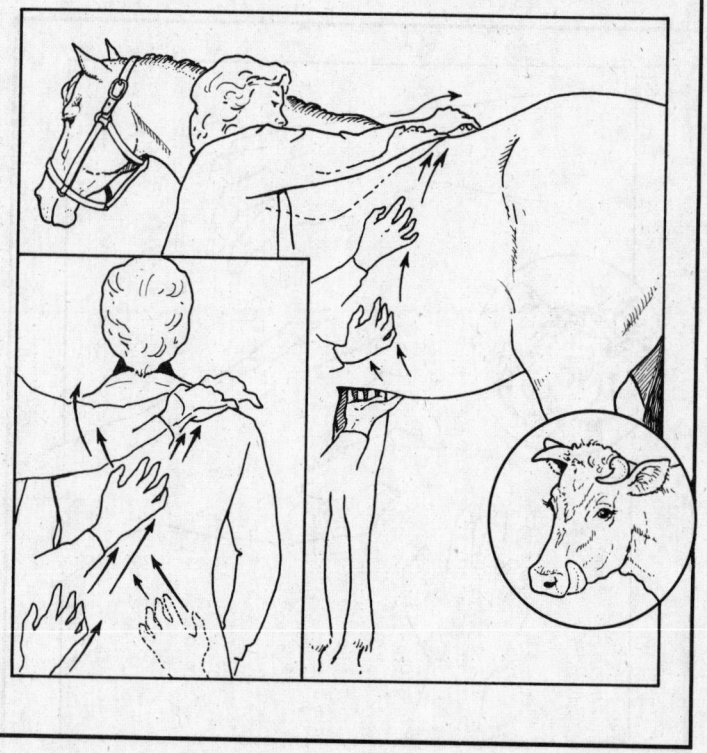

THE TIGER TTOUCH

When: The Tiger TTouch was named one day when I was demonstrating the hand positions on a friend. He was heavily muscled, and accustomed to very deep body work, so I spread my curved fingers apart and went straight in, leading with my nails. Looking at my hand, I was suddenly reminded of the shape and power of a tiger's paw.

This TTouch is helpful not only for heavily muscled individuals but for itch relief.

How: Hold the fingers raised and apart, curved like a tiger's paw, and use a pressure of a six or seven to make your circles. In this separated position, all the fingers and fingernails seem to make their own individual circles.

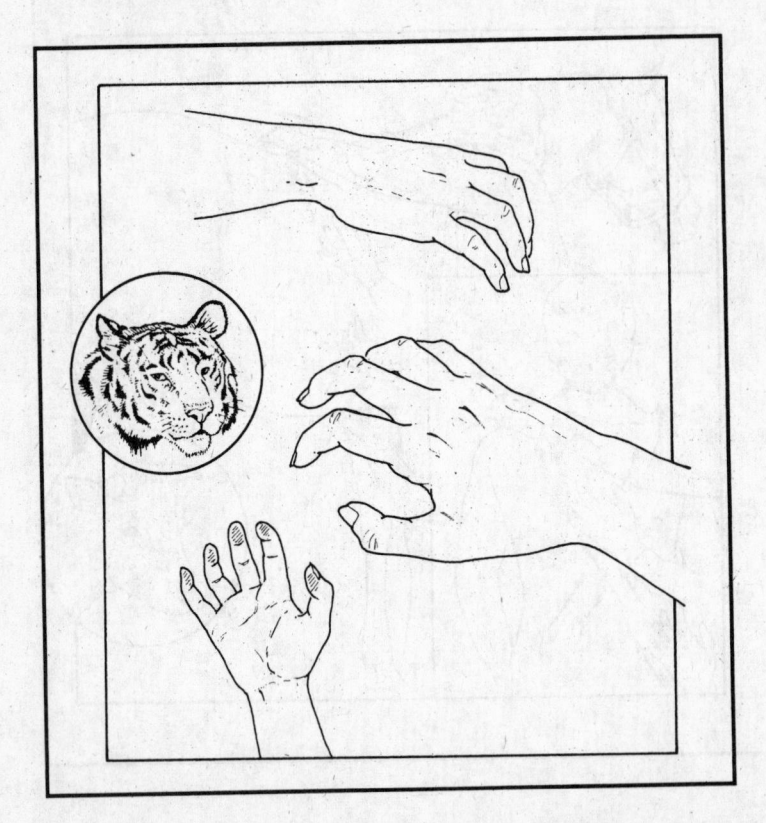

Noah's march

When: This is what we call the long, firm strokes of the hand all over the body with which we close a TTouch session. After the experience of revivification that the TTouch has brought to individual parts of the body, Noah's March brings back a sense of wholeness and reintegration.

How: Begin at the head and make firm, long strokes. Cover every inch of the body.

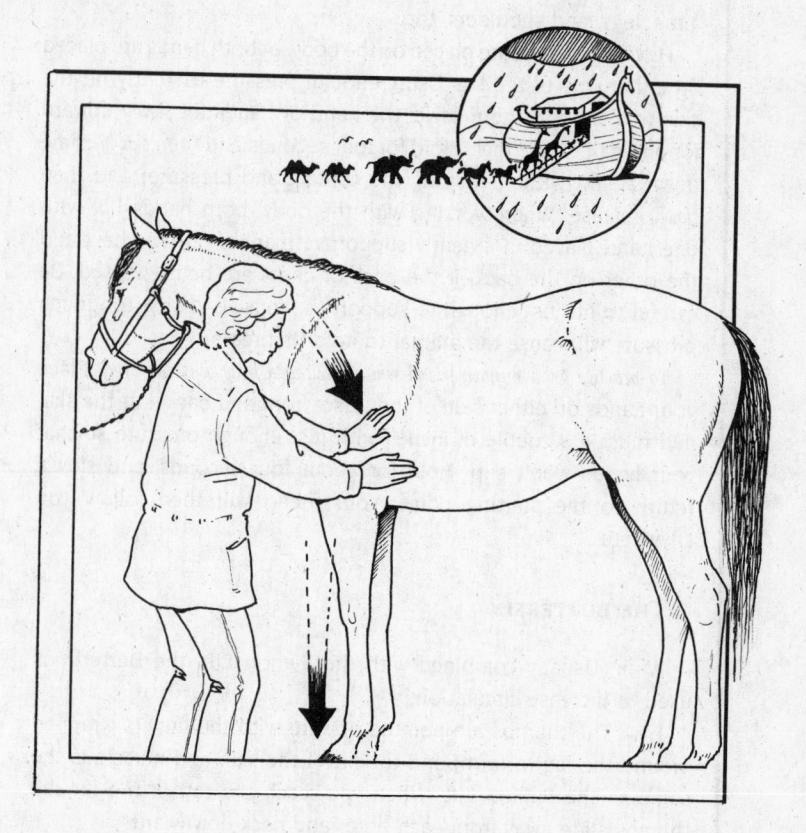

THE PYTHON LIFT

When: We named this position after the Burmese python Joyce (page 146), and it is a "lift" we use frequently to relieve and release muscular tension and spasm. It is effective on the shoulder, legs, neck, and chest areas of animals, and is great on human backs, arms, legs, and shoulders, too.

How: Both hands are placed on the body, or both hands are placed on either side of the leg. Using enough pressure to gently lift the skin and support the muscle, the hand or hands lift *slowly* upward for one-half to one inch. Hold for four seconds and then *slowly* come down again (without varying the contact and pressure) and then *slowly* release. When working with the body, both hands lift, with one hand placed in a gently supportive manner on the chest and the other on the back, if the back or chest are being worked. Be careful to lift just enough to support the muscle lightly; too much pressure will cause the animal to hold its breath.

To practice on a human friend who is feeling a little tense or tired: Place your hands on either side of the torso, front and back. Lift the skin and muscle a couple of inches with just enough pressure so that your hands won't slip. Hold for about four seconds, and slowly return to the starting point. Your friend will then follow you anywhere.

THE BUTTERFLY

When: Usually combined with the Python Lift, the Butterfly is used to increase circulation.

How: The thumbs are pointed upward with the fingers wrapped around the leg (or arm) and the lift of the skin and muscle is the same as the Python Lift. When you reach the top of the lift the thumbs slide away from each other and back downward.

TARANTULAS PULLING THE PLOW

When: This movement was inspired by a supposedly ancient Mongolian treatment called "skin rolling," which was employed to release fear before battle. Known as *chua'ka*, it was said to increase circulation and break habitual patterns of emotional response by releasing skin that had become frozen to muscle because of long-standing holding patterns. While we found *chua'ka* too painful for most animals (and people), the procedure did inspire us to invent a more pleasant and less invasive version of skin rolling.

How: This movement is done with or across the grain of the hair growth rather than against it. Place your hands side by side with the fingertips separated and curved in the way you might imagine two big spiders. The thumbs, lightly touching each other, extend behind the fingers. You then "walk" the forefingers and middle fingers of both hands simultaneously forward while allowing the two thumbs to follow behind, pushing a light furrow of skin along ahead of them like a plow.

My walking fingers reminded me of a very sweet pet tarantula who lived with me years ago in my office at the Pacific Coast Equestrian Research Farm. I named her Charlotte. (I had no idea how to tell the sex of a tarantula but I wanted to name it after the wonderful spider in E. B. White's *Charlotte's Web*.)

Animals who won't tolerate stroking, brushing, or being touched in any way find the skin rolling of Tarantulas Pulling the Plow a comforting introduction to the TTouch circles. It's also very effective with older animals who are sensitive when touched. The long lines have the effect of connecting the whole body horizontally from shoulders to hindquarters.

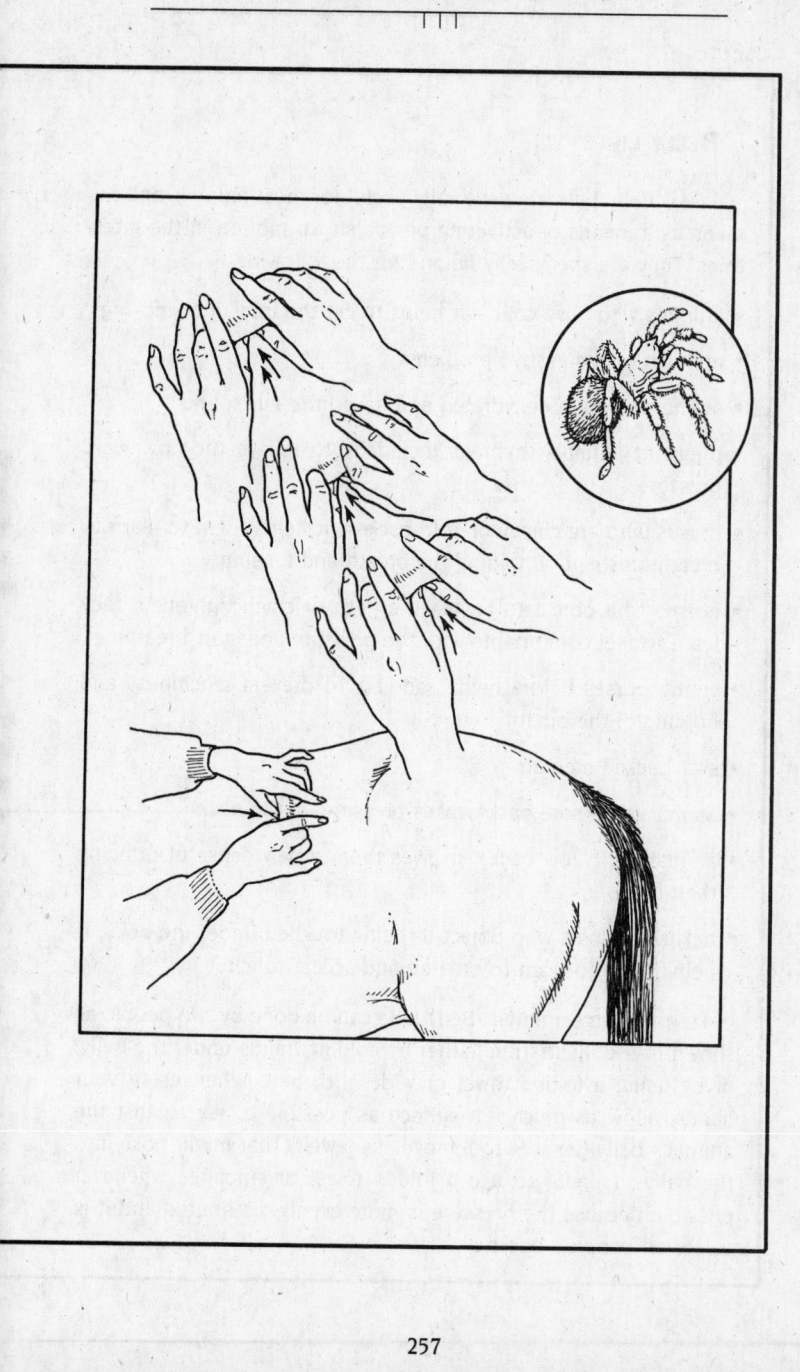

BELLY LIFTS

When: Belly Lifts are especially useful for most animals and humans as a means of activating peristalsis, or motion in the intestines. They are specifically helpful for the following:

- animals who have colic—it helps to get the intestines moving

- animals with digestive problems

- animals who are dehydrated and have little gut sound

- pregnant animals (humans included), to relieve the downward pressure

- horses who are cinchy or cold-backed; it helps them to learn to breathe instead of holding the breath and tensing

- horses who object to having the girth or cinch tightened; they learn to feel comfortable with the pressure and start breathing

- young horses before being saddled, to prevent cinchiness and holding of the breath

- swaybacked animals

- animals with sore backs; takes pressure off the back

- animals with low backs; it gives them a new sense of bringing their backs up

- ticklish animals who object to being touched under the belly; it helps them to learn to breathe and accept touch

How: On large animals, Belly Lifts can be done by two people as shown in the illustration, either by holding hands under the belly, or by using a folded towel or wide girth belt. When using your hands, allow as much flat surface as possible to lay against the animal's belly (be sure to remove any jewelry that might poke into the belly). I prefer to use a folded towel or surcingle whenever possible because the pressure is more evenly distributed and it is easier on people's backs.

Starting just behind the front legs, gently lift the animal's abdomen. Hold that position anywhere from ten to fifteen seconds, depending on the reaction. Then *slowly* release the pressure—the slow release is of utmost importance in getting the desired effect. If you can make the release twice as long as the lift, that would be ideal. Be sure to use your legs, not just your back to lift. If your animal objects, lift until you can just feel the downward pressure of the belly. Move three to six inches toward the hindquarters and repeat the procedure. Continue until you are as close to the flank as seems comfortable and safe (some animals are very ticklish or sensitive in this area, especially when in pain). The Belly Lifts can then be repeated three or four times, starting each time up toward the elbow.

When you're alone you can use your forearms and hands to do a Belly Lift. Again, be sure to use your legs when you lift, rather than your back and shoulders.

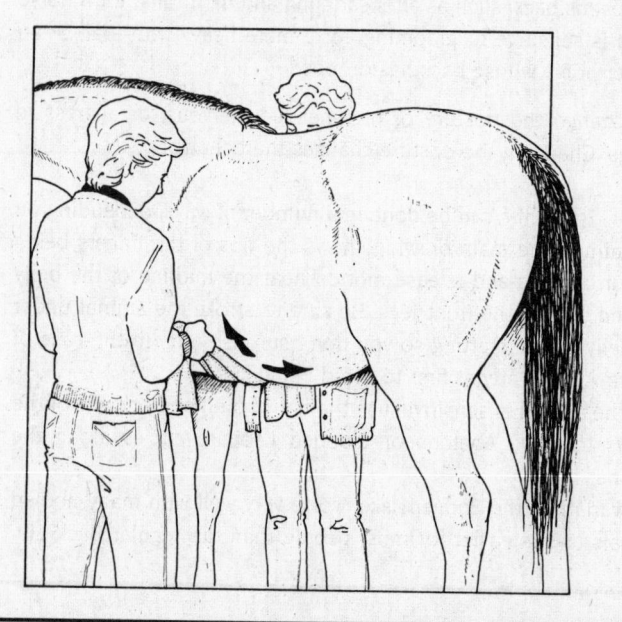

BACK LIFTS

We are often asked what the difference is between the Belly Lift and the Back Lift. The major difference is that with Belly Lifts the intention is to get the abdominal muscles to relax, allowing for deeper breathing, while with Back Lifts you are asking the animal to lift the back. The Back Lift changes the relationship of the vertebrae and allows the animal to lower and lengthen the neck.

When: Back Lifts are useful in the following cases:

- to lower the head

- to give the animal a way of activating the belly muscles and experiencing a new feeling in the back

- for swaybacks

- to fill out hollow areas in the back and withers whenever a horse drops his back, such as after saddling and mounting; for a horse who is sensitive to grooming, for a mare heavy with foal, or an older horse whose back has dropped

- to change the posture of high-headed, nervous, or aggressive dogs. Changing the posture changes the behavior.

How: Back Lifts can be done in a number of ways, depending on the animal. The main drawing shows the tips of the fingers being used in a press-and-release motion near the midline of the belly starting behind the front legs. Be sure to stroke the animal under the belly before starting so you don't surprise him. In the case of horses, start gently at first to avoid being kicked.

If the animal is sensitive, use the flat of the hand or pads of the fingers to make Abalone or Clouded Leopard circles along the midline.

A variation, the Badger Rake, works very well with many hoofed animals and is easier for most people than the regular Back Lift.

With your fingers apart and curved upward, start on the far side of the belly midline. Using a raking motion, bring both hands across the belly and partway up the barrel of the body. You can start out gently and gradually increase the pressure if the animal is not responding. In most cases it simply takes practice to become proficient in raising the back, but there are some animals who are extremely "stuck" in the back, and it can take several sessions to get them to move. It's useful when doing the Rake to have someone there to tell you what's happening, since in this position you are slightly bent over and cannot see whether the top of the back is rising.

When using the Badger Rake you will often see the back expand, making the animal look as though he has gained weight.

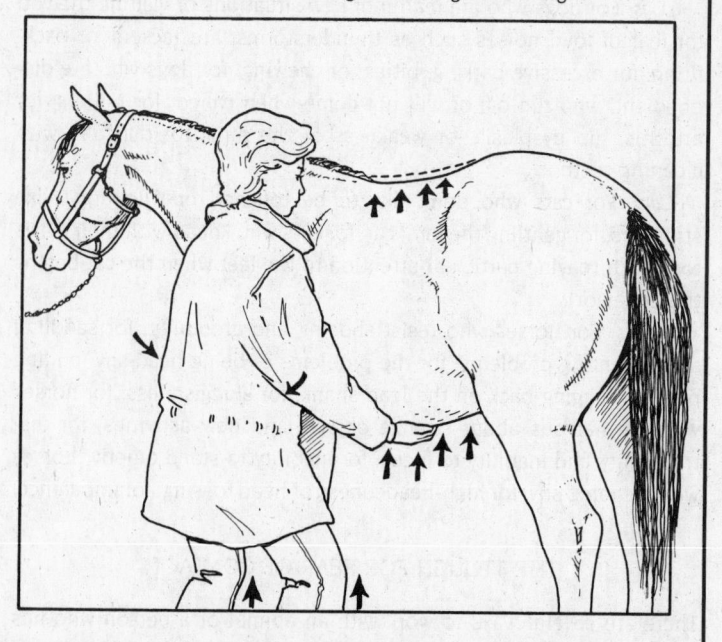

PHYSICAL AND BEHAVIORAL PROBLEMS THAT CAN BE AIDED THROUGH TTOUCH

General uses for a wide variety of animals: For shock due to injury or other causes; to ease resistance to veterinary examination; to ease fear/pain reactions and prevent panic injuries during restraint; to aid and speed recovery from serious illness, injury, or neurological damage; for colic; for reproductive difficulties; for neonatal problems; to relieve the stress of confinement; for neurotic behavior due to stress; for aggressive behavior toward owner, trainer, or handler; to relieve the stress of boredom; to improve social skills; to relieve the problems of aging; to calm hyperactivity; for training animals who are used in pet therapy. (We have used the TTouch successfully in preparing animals to enjoy contact with people in hospitals and rest homes. The people are then instructed to use the Lying Leopard TTouch on their animal guests— delightful results ensue for all.)

Dogs: For dogs who are fearful of new situations or visiting the vet; for fear of loud noises such as thunderstorms, firecrackers, or back-firing; for excessive barking, biting, or chewing; for dogs who are dis-obedient, who run off, or will not come when called; for pullers; for arthritis, hip dysplasia, or weakness of the hips; for difficulty with cleaning teeth.

Cats: For cats who don't like to be touched or who don't like strangers; for gentling the feral cat; for cats who knead with their claws too much (paying particular attention to the feet when the cat begins the behavior).

Horses: For horses who resist shoeing and grooming; for saddling and mounting problems; for the problems of being head shy, pulling back, or lugging back on the lead shank; for sluggishness; for horses who are nervous about strange objects and new activities; for dis-tractibility and inability to focus; for inability to stand quietly; horses who are ramp shy; for high-headedness or head tossing; for imbalance.

THE TTOUCH FOR FEAR OF CONTACT

There are several ways to work with an animal or a person who has difficulty being touched. To begin with, turn your hand so that you are approaching with the back of the hand rather than the front. Not only does the back of the hand present less of a threat in terms of the possibility of grabbing and holding, it also seems to emit less personal energy than the palm side.

To introduce the TTouch when the animal is fearful, make the initial circles fast, light, and widely separated. Make the first six to eight circles open three-quarter circles, being sure that the six o'clock position where you begin the circle is at a point parallel to the ground. Then as you feel the animal begin to accept your touch, slow down and go back to the regular TTouch of a circle and a quarter.

If even the lightest circles cannot be tolerated, switch to the Feather TTouch. Imagining you have feathers at the ends of your fingertips, barely flick the body and then swiftly pull away, and again brush and pull away. The contact is so light and short that the nervous system has no opportunity to respond habitually and will very quickly release the fear of contact. After "feathering," follow up with three or four three-quarter circles and then very light Clouded Leopard TTouches.

Sometimes, for a variety of reasons, we find that an animal or person can't handle any kind of human touch at all. When this happens I call on my collection of stuffed animals for help. A large, soft toy animal with big paws is preferable. Take one of the toy's paws and use it to give support and keep contact. With the other paw, make circles on the animal just as though the paw was your own hand. You'll find that when the intensity of the direct contact is removed, the animal or person who is afraid will be able to relax and even enjoy the TTouch. And just to make things even better, it's fun.

TTEAM TERMS

Body Wrap:

For Dogs: The body wrap is often helpful with large dogs who are hyperactive and won't stay still for the TTouch; dogs who are difficult to lead because they pull on the leash; shy dogs who tuck their tails between their legs or are afraid of thunder or loud noises; dogs who get "stuck" or "frozen" when they are being led; dogs who are being trained to wear a harness or backpack. It helps rebalance and settle a dog both physically and emotionally.

When leading a dog who pulls or freezes, we use a regular collar

combined with one of the dog halters (not a choke collar—see Halti, page 265). The figure-eight body wrap seems to improve the connections throughout the bodies of many large dogs, but its application for smaller dogs is more difficult because they can wriggle out of the configuration more easily. In such cases, combine a dog halter with a small rope tied around the hindquarters and attached to the harness.

For most dogs, we use a seven-foot-long, half-inch-diameter wrap, or one of an appropriate length and weight for your dog if he or she is small.

The body wrap gives the dog a sense of connection throughout his body from shoulders and chest to back and hindquarters, which is absent in a dog who is fearful, uncertain of his environment, or lacking in confidence.

For Horses: As with dogs, we use the body wrap to give horses a sense of complete connection from shoulders to hindquarters. It has the effect of settling nervous horses and improving their balance, coordination, and gait.

Chunking: The term, which we took from the language of neurolinguistic training, means the breaking down of a lesson into individual small "chunks" or steps in which an animal can be successful. Each small segment successfully completed builds confidence, makes the next step easier to learn, and eventually adds up to an entire lesson successfully executed. In other words, if you ask an animal to do something that it seems it can't do, instead of forcing the issue, go back a couple of steps in the lesson and ask the animal to perform a very small part of the lesson, perhaps in some other way than you have been attempting. Even if the animal is seemingly resistant, assume that it's a question of not understanding, and give the animal a chance to perform successfully.

Ear Work: The first time I successfully used ear work in the case of a life-threatening colic was 1958, although at that time I had no idea of the potential of the work. It wasn't until 1981 that I began to use the ear stroking from base to tip for cases of colic, shock due to injury or illness, fatigue, hypothermia, and the problems associated with aging.

Over the past decade, we have used the ear work for horses, dogs, cats, goats, sheep, cows, donkeys, bears, llamas, cheetahs, snow leopards, birds (around the ear orifice), and humans.

According to acupuncture theory, the ear is like a microcosm of the body, which may explain why this work is so effective.

How: Holding the ear gently between thumb and folded forefinger, slide and stroke upward from the base to the very tip, covering all parts of the ear. Make sure to include the tip, where a point for shock is located. In the case of tiny animals, slide the ear between the tips of your forefinger and thumb. When working the ear always think of it as having the delicacy of a rose petal.

In case of injury it's useful to do the Raccoon circles all over the inside and outside of the ear and then finish with the long strokes.

Halti: A training aid for boisterous or strong puppies, as well as a problem solver for "pulling" or hyperactive dogs of any age. With the Halti, when the dog behaves, he is free to pant, yawn, or "talk." If he pulls, struggles, or tries to bite, the Halti closes his mouth and diverts his head, preventing the problem. When used in combination with a leash on the collar, control is painless, simple, and creates respect for the leash.

How: To prevent resistance and fear of pressure, carefully prepare the dog before the halter is put on the head by using Raccoon TTouches all over the muzzle. Never use a leash on the Halti alone. Instead, use two leashes, one attached to the collar and one to the Halti. Great care must be taken not to pull or jerk on the Halti in any way that could put undue pressure on the muzzle and injure the dog. After a few lessons the Halti can be removed and the dog will have a much greater respect for the leash and will understand the signals on the collar.

The Halti is meant to be used as a kind method of training, never as a punishment.

It is interchangeable with the "Gentle Leader."

Mouth Work: Working the mouth of an animal is one of the fastest and most effective ways of making permanent changes in resistant animals and animals with emotional difficulties. We have successfully used this technique on many species from domestic to exotic, from horses, cats, dogs, and llamas to rhinoceroses, tigers, monkeys, and cheetahs.

We find that one to two sessions of TTouch mouth work lasting five to ten minutes is enough to change the attitude of animals who have a tendency to bite, puppies who are in the chewing phase, or young horses who are mouthy.

How: Speaking in broadly simplified terms, the mouth is connected neurologically to the limbic system, which in turn governs emotion. Therefore, working the mouth with small, gentle circles, sliding the fingers carefully over the gums and carefully working the lips both inside and out, can permanently change such difficulties as oversensitivity, anxiety, aggression, and fear-biting.

Caution: It is, of course, important to be cautious with aggressive dogs or other animals, and should you have any reservations about your safety, I strongly recommend that you begin by placing the animal in the Homing Pigeon or the Taming the Tiger positions so that he can be controlled and you can work him from a distance with the wand.

This gives you the opportunity to calm and prepare him and to safely see at what point he is ready to accept the TTouch of your hand. After the animal is in position, tie something soft over the knob end of the wand and gently use that to work the circles up over the head and to the lips.

Make sure the head is down and the animal's breathing is calm. Remember your own breathing, and tone to the animal while you work with him.

Obstacles: The ground poles (four to six inches thick and twelve feet long), PVC pipe (plastic plumbing pipe, which can be used instead of poles), cones, ramps, rubber tires, and plastic that are used in various ways to change habitual movements and attitudes. These are arranged to form different patterns of obstruction for working the animals.

Raised poles in various patterns are not only effective for improving coordination, agility, concentration, and focus, but can also be used to aid animals in healing after neurological damage. For example, with Kenyon the bear (page 225) we placed the poles in parallel lines about eight inches high so that in stepping over them it was necessary for Kenyon to use his back and shoulders in a nonhabitual way.

Plastic PVC pipe three inches in diameter and cut in six-foot lengths can be used for dogs. They are easy to store and you can lengthen them by joining them together with plastic sleeves that are sold for that purpose.

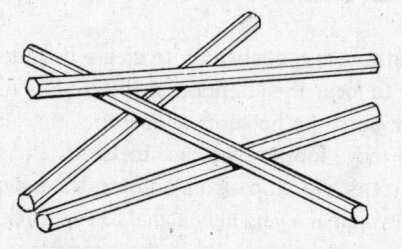

Pick-up Sticks: An exercise in mindfulness for both horses and dogs.

For Horses: The poles are arranged just as the name indicates, in a pattern of obstacles that encourages horses who are easily distracted to pay attention to the handler.

For Dogs: The plastic or wooden poles are laid across each other randomly in such a way that the dog has to pick his way through them. He has to stop and think rather than rush through them. He is not able to sit down between them but rather must use his mind quietly to figure out how to get through.

Labyrinth: A maze set up in different patterns with a minimum of six twelve-foot poles. It is used to teach cooperation and partnership to both animal and handler, and as a means for animals, especially horses and dogs, to learn obedience, balance, self-control, focus, patience, and precision. The maze is also used for rehabilitation when there has been neurological damage that impairs movement (page 56).

For dogs: In constructing a labyrinth for dogs you can use plastic PVC pipes joined together, bamboo poles, or long rods. The poles or obstacles should be about three to four inches thick.

For cats or small animals you can make a maze using pencils, string, or sticks.

I first began to use the labyrinth to create boundaries that would provide a way to focus the attention of a horse in a nonhabitual way. To begin work, lead the horse or animal into the labyrinth and get him or her to move forward two or three steps and then stop, then two or three steps and stop again, and the same again. To do this, simply give the animal a very light signal to stay back, either with the chain (in the case of a horse), with a slight movement of the wand in front of the nose, or a light tap on the chest. These movements, which teach the animal to hold back, to shorten the step, and to inhibit movement, create balance, focus, and concentration.

Once the animal can comfortably accomplish the step-and-stop pattern, we go on to the half-walk, in which we ask him to shorten his step, to inhibit his own forward movement, and to hold himself back, the object being to learn self-control. When we don't do this but simply take the animal through the labyrinth and around the corners at a regular walk, we find that though it teaches the animal to pay some attention, he quickly gets bored. The use of light signals to stop the animal in a balanced position and to shorten steps actually prevents boredom, and in a biofeedback experiment with horses we even discovered that when horses shorten their stride to round a corner, the beta wave is activated, whereas if they are walking at a normal pace in a straight line beta is not activated. (Beta brain waves are those in evidence during thinking and problem solving.)

The eventual goal on the turns is to have the animal move around a turn in such a way that the hind feet follow the same path as the front feet and the rhythm of the movement continues smoothly and uninterruptedly. Nervous animals will often stop at the corner and take two or three steps with the front feet before they move the hind feet, indicating a lack of coordination in the slowed movement. By the fourth time through the labyrinth this tendency usually disappears.

Working within the boundaries of the labyrinth seems to calm and focus dogs as well as horses, and animals that have a tendency to be hyperactive quiet almost immediately. When working with cats or smaller animals you can guide them through the labyrinth you've made using the eraser end of an unsharpened pencil as a wand.

Horse Lead: A thirty-inch chain sewn to a six-foot soft nylon lead that we attach to the halter in a specific manner so that the chain goes over the noseband of the halter. This encourages the horse to lower his head, which overrides the flight instinct and allows the handler to give clear, precise, subtle signals without using force. The lead is to be used only as a light signal when combined with the movement of the wand, not to shank, jerk, or pull. The lead is available from the TTEAM office in New Mexico (see page 277).

Tail Work: We have found tail work to have many uses. Working on the tail with gentle pull-hold-and-release movements, together with circles around the base, underside, and top of the tail strengthens the back and hindquarters and relieves stiffness there. The work also seems to release fear, especially for animals like horses and dogs, who have a tendency to clamp the tail when frightened.

We have found that altering the way the animal holds the tail changes his or her response to fearful situations. For example, we have had impressive positive results in innumerable cases when we have used the tail work to alleviate such problems as fear of loud noises (backfire or thunder), fear-biting, aggression, or timidity.

With hoofed animals we have had similar success with tail work in reducing kicking.

For nervous horses and dogs, tail work is helpful in assuaging fear of new situations, and to assist breeding or palpation for breeding by the veterinarian. In fact, working the base and top of the tail of bitches or mares greatly facilitates ease of breeding in maiden or nervous animals.

For pups who have docked tails it is helpful to do very gentle circles around the tail to relieve nervousness and imbalance.

Toning: To tone is to use the voice to calm and quiet the nervous system and breathing of both the animal and the human who are working together. Words are drawn out into a soft, continuous sound, i.e., "gooood," "eeeeasy," and "staaaand." This way of communicating dispels tension and grounds both handler and animal.

The long descending notes create an atmosphere of calm and confidence, as opposed to short, sharp words, which demand instant obedience and which cause an animal to hold the breath and freeze the motion. Toning is a positive training aid in educating an animal, as opposed to training through the use of fear.

Wand: The TTEAM wand is a whip that is four feet long (three feet for dogs), strong, light, and well balanced—neither too stiff nor too flexible. All TTEAM wands are white because that seems to have a positive visual effect on animals. They are tipped with a small, hard, buttonlike knob that is valuable for making circles from a distance.

Since the word *whip* carries the connotation of punishment for most people, we use *wand* in order to convey the TTEAM spirit in which it is employed. We use the wand as an extension of the arm—to stroke an animal in order to calm and focus it, to give reassurance, and to convey signals. (Sometimes this light stroking quiets an animal so fast it *does* seem like magic.) The wand is available through the TTEAM office (see page 277).

THREE TTEAM POSITIONS

The Dingo: The position we use to teach an animal to lead and go forward easily when asked.

Hold the wand in your right hand (when you are on the left side of the animal). Hold the lead in your left hand, a few inches from the halter when working with horses or hoofed animals, or the equivalent when working with dogs or smaller animals. Stroke the wand along the length of the back in a two-stroke motion, and then give the "come forward" signal with a short pull-and-release on the lead, reinforcing the signal with a double tap-tap of the wand on the top of the hindquarters.

The long strokes of the wand along the back steady the animal and connect him with a sense of his body as a whole.

The Dingo is useful in teaching animals who have problems with leading, whether they pull backward or simply stand stubbornly in place.

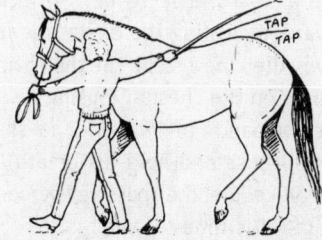

Journey of the **Homing Pigeon***:* For *dogs and* hoofed *animals:* A position in which two handlers are used, one on either side of the animal.

This position, by dividing the animal's attention, breaks habitual response patterns and greatly accelerates the ability to learn. Animals worked in this position are much quicker to cooperate than when they are worked by a single person from one side.

For *horses:* This position quickly teaches a horse to override the flight instinct, focus, and become more obedient and balanced. It teaches the handler to lead from the off side so the horse becomes accustomed to having a person on both sides.

For *dogs:* The Homing Pigeon is particularly useful for dogs who are timid, nervous, afraid of noises, aggressive, or fear-biters (dogs who snap or bite strangers out of nervousness). This position permits you to stroke them from a distance, and once you have calmed them and won their trust you can then move to the TTouch body work.

How: Fasten a lead to each side of the animal's collar or halter. Each handler holds the lines in both hands far enough away from the animal's head to avoid giving the animal a sense of claustrophobia—with horses and dogs approximately two to four feet away, depending on the behavior of the animal. The wand is held in the left hand.

To go forward, one handler first strokes the animal on the chest with the wand, then moves forward. Holding the wand about two feet in front of the animal's nose so he can focus on it, she sweeps it smoothly forward, indicating the direct line in which she wishes the animal to move. After a few steps she then brings the animal to a stop by tapping lightly on the chest and using a softly intoned word command such as "staaand" or "whoooa" (a sharp command tends to make the animal tense and hold his breath). The two handlers take turns with this procedure, the "passive" person moving along with the "active" one until it is time to switch.

If the animal is reluctant to move forward after the initial movements of the "active" person, the "passive" person can encourage him forward with a stroke on the back and a tap on the hindquarters.

Note: Before starting the Homing Pigeon, you must decide which one of you will take control should the animal become upset. It is usually the person with the most experience. In the case of horses it can also be the handler on the left side since most people have more experience on that side. It's also important to communicate with your partner so you both know when you will be turning, stopping, and so forth.

Taming the Tiger is useful to owners, groomers, handlers, and veterinarians who wish to work on an animal that is too nervous, fearful, or aggressive to be handled without restraint. An excellent method for teaching an animal to stand quietly.

It is particularly appropriate when a person is working alone, as a means of safely and nonviolently restraining an animal that has a tendency to bite, pull back, kick, or rear.

How: Place your animal a few feet from a post, or a wall into which you have placed a ring. Attach a lead to the animal's collar or halter on one side. On the other side, attach approximately eight to twelve feet of rope, depending on the size and species of animal, and take it around the post or through the ring. Then thread the end of that rope back through the collar or halter, under the animal's chin, and back to the same hand in which you are holding the other lead line. Hold both the rope and the lead line with a finger separating them.

This gives you control of the animal's movements; you can keep a light contact in order to restrain the animal, and then, when the animal

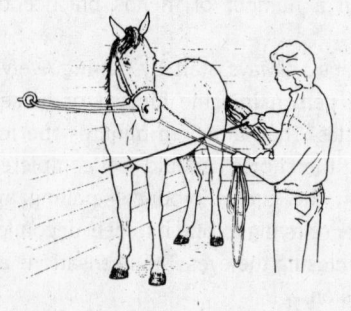

273

begins to relax and cooperate, you can slacken the pressure on the two lines as a reward. The object is to get the animal to accept the restriction of which she is so afraid. If the animal starts to pull back, use the Dingo to stroke and tap her forward.

Note: This position can also be useful in preparing a fearful show dog to calmly accept the judge's examination.

LEARNING

People often ask me how it is that I can work with whatever creature is presented to me—a tiger, a horse, a snake, a tiny bird—so many different animals, so many different species. The wonderful diversity of life seems as though it would call for correspondingly diverse responses, but actually what makes the TTouch communication possible is not attention to what separates but recognition of what is shared.

As I enter the particular world of each animal, my intent is to approach noninvasively. Instead of seeing the TTouch as something that I do *to* animals, which would create separation between us, I view the circles as a way to come into cellular harmony with them, a way of allowing my cells to speak to theirs. At a cellular level, no living thing is alien to any other, and so the sense of connection remains the same whether I'm working with a gerbil or a lynx, a kitten or an elephant.

Of course once the communication is established you want to be able to "listen" and fine tune your contact. Communication being what it is—a two-way proposition—you'll need to know not only how it feels to give the circular TTouch but what it's like to receive it. We found that for the first lessons, the variety of experience and feedback that comes when a number of friends practice on each other is invaluable.

In my workshops I always start by pairing everyone off and suggesting that one person imagine himself or herself to be a horse receiving the circles. Then after ten minutes the roles are reversed. The person being the "horse" should feel completely free to act out whatever reaction the circles produce—pulling away, backing off, threatening to kick or trying to bite if it feels uncomfortable or painful, and relaxing and closing the eyes if the sensations are pleasurable or are releasing tension.

More than likely, what you'll discover is how surprisingly different everyone is. Some people are sensitive to the lightest whisper of a touch while others respond only to deep and heavy pressure. Learning this, you begin to appreciate that animals have the same variety of experience and range of feeling and that exploring a creature with the TTouch can be a key to understanding behavior as well.

FINDING THE RIGHT PRESSURE

A number three to four pressure is the best way to start out a TTEAM exploration, whether on an animal or a human. From there you can go in either direction in the pressure scale to discover the range that is most suitable for the recipient. Say you are playing the role of the "horse" and as I work on your shoulder with a number two pressure you swiftly move away. That would indicate to me that perhaps you are overly sensitive there because of an old injury, or patterns of holding the shoulder, or because of an inflammation.

My response would be to put my hands on you in a way that would help you to overcome that sensitivity and fear. Still using the same one or two pressure, I would move into the Lying Leopard TTouch to give you the reassuring warmth of my hand, and the Python Lift to ease and regulate your breathing. Normally, as fear and discomfort disappear, a number five will become comfortable and even enjoyable.

Or take a case at the opposite end of the scale—let's say even though I increase pressure from a five all the way up to a nine or ten, your response remains sluggish and minimal. The appropriate step to take in this case would be to switch to the Bear TTouch, which allows the fingers to penetrate more deeply into the areas of the body that are insensitive, yet does not demand the exertion of heavy pressure. With this penetrating TTouch you are actually giving the cells an opportunity to *feel* and the nervous system a chance to *respond*, thereby *increasing* the animal's sensitivity.

Usually, for humans practicing on each other, I suggest working on only one side of the body. This allows a basis for useful "before and after" comparisons. For instance, you can compare how much lighter and more mobile your arm feels on the side that has been worked with the Python Lift than on the corresponding side. The session is completed with a discussion in which the person who has been playing the "horse" goes from nonverbal feedback to becoming a veritable Mister Ed and describing what he or she felt. Were the circles round enough? Were they consistently smooth all the way around? Were they too fast or too slow, and so forth.

When it's your turn to be the "horse," observe how you feel when something your partner is doing bothers you. Do you tighten up your muscles, restrict your breathing? Can you feel the difference when your partner is breathing into the circle and holding the hand softly, and when he or she is not?

When it's your turn to take the active human role, feel how the body of the "horse" changes when you respond sensitively to the signals you are receiving. See how the "horse" mirrors any tension you may be feeling. Talk to each other about your findings and before you know it the TTouch will have moved out of the theoretical world of the printed page to become your own personal experience to be shared and explored further with your dog, cat, horse, gerbil, snake, or bird.

RESOURCES

There are two Tellington TTouch Training offices, one in New Mexico and one in Canada. The former houses our international offices for TTouch trainings for companion animals, horses (TTEAM), and our TTouch work with humans called "TTouch for You," as well as our Animal Ambassador program. Here, Carol Lang and Kirsten Henry lead our staff in scheduling clinics, answering queries, managing the Web site, filling orders, and coordinating programs in twenty-seven countries. Our offices are housed in an adobe building with a view to distant pinkish arroyos. For many years this was also the rehabilitation home of six pigtail macaques who were visited by many schoolchildren over a ten-year period as part of our Animal Ambassador program.

This book is intended as a general overview and introduction to TTEAM and the Tellington TTouch. Please feel free to write us: we would like to hear how the TTouch works for your animal friends. Check our Web site at www.ttouch.com to find valuable information about how TTouch can help your favorite animals, whether they be dogs, cats, horses, bunnies, birds, or reptiles. There are stories about work with all species as well as a store where you can order our books, DVDs, and equipment, or receive a copy of our TTouch newsletter. You can find a certified TTouch practitioner in your area for personal help, or find a workshop you would like to attend.

At www.ttouch.com/forum you can read dozens of stories and get answers to your questions.

"Tellington TTouch Training International"
Linda Tellington-Jones
P.O. Box 3793
Santa Fe, New Mexico 87501
Phone: 1-800-854-TEAM
Fax: 505-455-7233
E-mail: Info@ttouch.com
Web site: www.ttouch.com

To subscribe to the TTouch Newsletter, order books or DVDs or to find trainings in Canada contact:

"Tellington TTouch Training Canada"
Robyn Hood
5435 Rochdell Road
Vernon, British Columbia V1B 3E8
Canada
Phone: 1-800-255-2336
Fax: 250-545-9116
E-mail: TTouch@shaw.ca
Web site: www.tteam-ttouch.ca